Application of Computational Techniques in Power Systems for Power Quality Improvement

This book explores the role of computational intelligence techniques in addressing power quality challenges in modern power systems. It examines the integration of multiple energy sources, renewable energy systems, and energy storage units while analyzing nonlinearities in power plants. The book introduces optimization methods, including machine learning, genetic algorithms, and neural networks, to enhance power system stability, reliability, and efficiency. It bridges theoretical insights with real-world applications in modern power networks.

- Explores the use of AI techniques, including genetic algorithms and fuzzy logic, for power system analysis and optimization.
- Examines power quality issues such as frequency deviations, tie-line power flow variations, and area control errors
- Discusses the tuning of secondary controllers, including PID controllers, using advanced optimization algorithms
- Addresses PQ challenges posed by renewable energy integration and energy storage units
- Covers real-time system monitoring and control for maintaining power quality under nonlinear operating conditions
- Provides a structured approach to analysing system behavior using mathematical modeling and time-domain simulations

This book serves as a valuable resource for researchers, professionals, and students in power system engineering, computational intelligence, and smart grid technologies.

Knowledge-Based Engineering for Innovation

Nilanjan Dey

Professor with the Techno International New Town, Kolkata, and visiting fellow of the University of Reading, UK

The series presents invaluable insights into knowledge-based engineering for innovation by investigating the role of emerging technologies like machine learning and artificial intelligence in enhancing knowledge management, data-driven research, knowledge-intensive applications, knowledge transfer, and related areas. By addressing the challenges and opportunities in this domain, the series aims to bridge theoretical and practical aspects through transdisciplinary approaches, highlighting the critical role of knowledge-driven research in today's academic and industrial landscapes. The series promotes the formation of knowledge-driven communities, particularly in the Information Technology sector, while also bridging diverse fields such as healthcare, education, social sciences, business, and environmental studies to this platform for examining strategies, tools, and best practices that support collaborative and impactful research.

Application of Computational Techniques in Power Systems for Power Quality Improvement
Jagatheesan Kaliannan, Boopathi Dhanasekaran, Sourav Samanta, Anand Baskaran, Kanendra Naidu

www.routledge.com/Knowledge-based-Engineering-for-Innovation/book-series/KBEI

Application of Computational Techniques in Power Systems for Power Quality Improvement

Jagatheesan Kaliannan,
Boopathi Dhanasekaran,
Sourav Samanta, Anand Baskaran,
Kanendra Naidu

CRC Press
Taylor & Francis Group
Boca Raton London New York

CRC Press is an imprint of the
Taylor & Francis Group, an **informa** business

A CHAPMAN & HALL BOOK

First edition published 2026
by CRC Press
2385 NW Executive Center Drive, Suite 320, Boca Raton FL 33431

and by CRC Press
4 Park Square, Milton Park, Abingdon, Oxon, OX14 4RN

CRC Press is an imprint of Taylor & Francis Group, LLC

ISBN: 978-1-041-10944-0 (hbk)
ISBN: 978-1-041-11697-4 (pbk)
ISBN: 978-1-003-66115-3 (ebk)

DOI: 10.1201/9781003661153

Typeset in Nimbus Roman font
by KnowledgeWorks Global Ltd.

Publisher's note: This book has been prepared from camera-ready copy provided by the authors

Contents

Preface

Nowadays, the interconnection of different power-generating systems increases due to an enormous amount of technical growth, industrial development, and modern technologies to achieve satisfying load demand. The Automatic Generation Control (AGC) in the power system handles the sudden load demand and the delivery of stipulated power with good quality in a sudden and continuously varied load period. The sudden load disturbance affects standalone power systems' stability and power qualities. To overcome this issue, proper design of the power system and suitable controller modeling are more crucial when non-linearities and boiler dynamics components are incorporated in the system. Load Frequency Control (LFC) has a substantial role in electric power systems in interconnected areas. Reliable manoeuvre of the power system necessitates the power balance between the system's associated losses and the total load demand of the power generation. Thus, the LFC is used in the power system to keep the frequency and tie-line power flow of the system within the limit during sudden load disturbance. The main problem in the interconnected power system is to reduce the damping oscillations in the system frequency; thus, the tie-line power flow deviations should be kept within the limit during sudden load demand. When damping oscillations exist in the system response for a long period of time without any adequate controller, it affects the system operation and quality of delivered power supply. For providing good quality power and stable power system operation, extensive research work has been carried out and proposed in the last few decades. Due to the massive development in industries and technology, the load demand value is changeable and cannot be predicted as it varies randomly. Several efforts are carried out based on effective optimization methods to realize numerous benefits and purposes for the power systems' operation control. Researchers have conducted different studies to solve the optimization problems to optimize the power system secondary controller parameters. Differential evolution (DE), Particle swarm optimization (PSO), Firefly Algorithm (FA), Genetic Algorithm (GA), and Ant Colony Optimization (ACO) are examples of optimization algorithms that can be included in the design of PID controller parameters for effective operation of the thermal power system. In the power systems, the Proportional-Integral (PI) and Proportional-Integral-Derivative (PID) controllers are used as secondary controllers. Consequently, this book includes different applications of the optimization techniques to design the PID controller for LFC of single area as well as multi-area interconnected thermal power systems with and without incorporating non-linearities and boiler dynamics effects.

Author Biography

Jagatheesan Kaliannan is a Professor in the Department of Electronics and Communication Engineering at M. Kumarasamy College of Engineering, Karur, Tamil Nadu, India. He has more than 14 years of teaching experience and 12 years of research experience. He is the author of the book "Bio-Inspired Algorithms in PID Controller Optimization" published by CRC Press, Taylor & Francis Group, New York. Additionally, he has published more than 85 research papers in various reputed journals, book chapters, and international/national conferences with reputed publishers.

His Google Scholar profile shows an h-index of 23 and an i10-index of 42, with 1,734 citations. His primary research interests include power system operation and control, stability analysis, digital controllers, and bio-inspired optimization techniques. He has also organized several national and international-level technical events such as faculty development programs (FDPs), seminars, and workshops. Furthermore, he has published patents and served as a co-coordinator for various academic councils and social activities.

Boopathi Dhanasekaran works as an Associate Professor, Department of Electrical and Electronics Engineering, Paavai Engineering College (Autonomous), Namakkal, Tamil Nadu. He completed B.E (Electrical and Electronics Engineering) in the year 2008, M.E (Applied Electronics) in the year 2011, and Ph.D. degree in the year 2021. He has 16 years of teaching and 13 years of research experience. His research interests are Power system optimization, Power quality, and Power system stability analysis. He has published more than 20 research articles in reputed journals, as book chapters, and through conferences. His h-index is 11, and his i10 index is 11 with more than 350 citations.

Sourav Samanta is currently working as Assistant Professor in the Department of Computer Science and Engineering at University Institute of Technology, The University of Burdwan, West Bengal, India. Before joining the University Institute of Technology, he worked as a lecturer at the Gobindpur Sephali Memorial Polytechnic, Guskara, Burdwan, West Bengal. He has completed Ph.D. on Application of Computer Vision in Smart Agriculture at the Indian Institute of Information Technology, Kalyani. He has completed M. Tech in Computer Science and Engineering from JIS College of Engineering, WBUT, West Bengal and completed B.E in Information Technology from University Institute of Technology, The University of Burdwan, Burdwan, West Bengal respectively. He has approximately twelve years of academic experience. He has also completed a certificate program in Intellectual Property Rights and Technology Business Management from MAKAUT.

He is working with researchers from five different countries. His research area includes Nature Inspired Computing, Quantum Machine Learning, Computer Vision in Agriculture and Information Security. He has authored 75 research papers, including articles in reputed journals, conference proceedings, and book chapters published by IEEE, Springer, Elsevier, and others. As per Google Scholar, he has 2109 citations with an h-index of 24. He is a member of IEEE, ACM, and other professional organizations. He is the co-author of the book Optimization of Watermarking in Biomedical Signal published by Lambert Academic Publishing. He is also co-editor of the two books: Design Frameworks for Wireless Networks and Architectural Wireless Networks Solutions and Security Issues published by Springer in Lecture Notes in Networks and Systems Series.

Anand Baskaran received his B.E. in Electrical and Electronics Engineering in 2001 from Government College of Engineering, Tirunelveli and M.E. in Power Systems Engineering from Annamalai University in 2002, and Ph.D. in Electrical Engineering from Anna University, Chennai, in 2011. Since 2003, he has been associated with Hindusthan College of Engineering and Technology, Coimbatore, Tamil Nadu, India. Currently he is working as a Professor and Head, Department of Electronics and Instrumentation Engineering in the same institution. He is the Chair of the annual IEEE – International Conference on Sustainable Computing and Smart Systems (ICSCSS) being organized by this institution.

He received a research grant for a solar-related project from the Anusandhan National Research Foundation (ANRF), Government of India, as an ongoing research work. His research interests include electrical power system operation, control, optimization, and application of AI in power system problems. He has published more than 140 papers in national/international journals and conferences. He is an active member of IEEE, SSI, ISTE, and a Fellow in IE(I).

Kanendra Naidu is a Senior Lecturer at the Faculty of Electrical Engineering, Universiti Teknologi MARA (UiTM), Shah Alam, Malaysia. He holds both an M.Eng.Sc. and a Ph.D. in Electrical Engineering from the University of Malaya. His research focuses on artificial intelligence applications in power systems, including areas such as load frequency control, energy management systems, virtual power plants, and electric vehicle integration. He has authored over 60 publications in esteemed journals like the International Journal of Electrical Power & Energy Systems, IET Generation, Transmission & Distribution, and IEEE Access, amassing more than 2,602 citations with an h-index of 22. He has led and collaborated on various research projects, notably the development of Virtual Power Plants (VPP) for frequency control ancillary services, modeling of electric vehicle (EV) charging patterns integrating photovoltaic-based stations, and a project on multi-energy ship microgrids. Beyond academia, he consults on initiatives related to EV battery recycling, EVSE scheduling, and energy storage systems. Professionally, he is a Chartered Engineer with the IET (UK), a Senior Member of IEEE, and an active member of the Board of Engineers Malaysia (BEM) and the Malaysia Board of Technologists (MBOT).

1 Introduction

Computational intelligence (CI) is a branch of artificial intelligence (AI) that comprises systems that mimic humans for problem solving through the utilization of computational approaches. CI methods include neural networks, evolutionary algorithms, and fuzzy logic control systems. The main objective of CI is to solve complex real-world problems that cannot be addressed by traditional algorithms, such as optimization techniques and pattern recognition approaches. CI methods are adaptive and learn from data or experience rather than relying on predetermined rule-based data. Flexibility, adaptability, and learning capability are key features of CI.

Classification, regression, and pattern recognition are inspired by the human brain and are implemented through neural networks. The fuzzy logic method addresses incomplete or uncertain information in an environment, enabling reasoning under imprecision. Evolutionary algorithms mimic biological evolution to find optimal or near-optimal solutions for complex problems.

A hybrid CI approach combines multiple CI techniques, such as neuro-fuzzy systems, to achieve better optimization results. CI is widely applied in various fields, including robotics, image processing, finance, medical diagnosis, and natural language processing. The major advantages and characteristics of CI include its ability to address nonlinear, high-dimensional, and dynamic problems in real-world applications. With continuous advancements in AI, CI continues to expand its capabilities across multiple domains, helping to overcome critical challenges.

1.1 COMPUTATIONAL INTELLIGENCE

Load frequency control (LFC) for power systems, which is essential for preserving the equilibrium between electrical power output and consumption, heavily relies on CI. By modifying the power generation from different plants, the LFC seeks to control the frequency of the power system and ensure that it stays within a specified range. In situations where conventional control techniques might not be sufficient because of system complexity, unpredictability, and nonlinearity, CI offers sophisticated methods for optimizing and improving the performance of LFC systems. The major CI techniques are artificial neural networks (ANNs), fuzzy logic control (FLC), and evolutionary computation techniques. The benefits of CI in LFC include flexibility, adaptability, and optimization. However, it also has some challenges, such as complexity and convergence issues. In this chapter, computational techniques are discussed in detail.

1.1.1 NEURAL NETWORKS

An ANN is a computational model inspired by the structure and function of the human brain, which is capable of learning and adapting to new data, making it a

DOI: 10.1201/9781003661153-1

powerful tool in the field of data analysis and machine learning, where it can be applied to a wide range of tasks, such as image recognition, natural language processing, and predictive analytics, allowing for the extraction of complex patterns and relationships within large datasets and enabling the development of intelligent systems that can make accurate predictions, classify objects, generate highly accurate models of real-world phenomena, and even create new, synthetic data that can be used to augment existing datasets, reducing the need for manual data collection and enabling the simulation of scenarios that would be highly valuable in various fields such as finance, healthcare, and climate modeling, where the ability to simulate complex systems and predict the outcomes of different scenarios can be a game changer, allowing experts to make more informed decisions, driven by data and insights that were previously inaccessible, and enabling them to respond more effectively to a wide range of situations, from responding to natural disasters and predicting market trends to personalized medicine and optimizing business processes, leading to increased efficiency, productivity, and competitiveness, while also opening up new opportunities for innovation and driving growth in industries.

Artificial neural networks have already demonstrated significant potential in various fields and are poised to revolutionize numerous aspects of our lives, from healthcare and education to finance and transportation, and they will undoubtedly continue to shape the future of industries and society as a whole, as researchers and developers continue to push the boundaries of what is possible with these powerful tools, driving innovation and breakthroughs that will have a profound impact on numerous areas of human life, fueling unprecedented growth, and accelerating the development of transformative technologies that will continue to reshape our world and improve the human experience, from advancements in medical diagnosis and personalized treatment plans to autonomous vehicles that redefine transportation systems, intelligent robots that transform manufacturing processes, and smart cities that optimize energy consumption, waste management, and public services, ultimately leading to more sustainable, efficient, and livable environments for generations to come, where the seamless integration of AI and the Internet of Things (IoT) gives rise to entirely new forms of interaction, collaboration, and problem-solving, enabling humans and machines to learn from each other, adapt to changing circumstances, make data-driven decisions that drive innovation, foster economic growth, and create unprecedented opportunities for social mobility, cultural enrichment, and environmental stewardship.

Therefore, a new era of unprecedented technological advancements that will forever change the fabric of our society, from the way we communicate to the manner in which we work, live, and interact with one another, and ANNs form the backbone of this revolution, enabling machines to perceive, process, and respond to complex patterns and data, as well as learn from experiences, making them capable of simulating human intelligence and decision-making capabilities that are applicable to a wide range of fields, including but not limited to healthcare, finance, education, transportation, and cyber security, among others, thereby facilitating a previously unimagined level of automation and subsequently propelling human civilization forward with profound implications for the future of work, creativity, innovation, and technological advancements that will redefine the boundaries of human health potential, leading to breakthroughs in fields such as medicine, where novel treatments and personalized healthcare are being developed at an unprecedented rate, transforming the lives of millions of people worldwide and enabling the medical community to tackle some of the most pressing and complex health challenges of our time, from rare genetic disorders and chronic diseases to infectious and mental health conditions, thus significantly increasing our chances.

An ANN is a computational model inspired by the structure and function of the human brain and is capable of learning and adapting to new data, making it a powerful tool in the fields of data analysis and machine learning. It can be applied to a wide range of tasks, such as image recognition, natural language processing, and predictive analytics, allowing for the extraction of complex patterns and relationships within large datasets while enabling the development of intelligent systems that can make accurate predictions, classify objects, and generate highly accurate models of real-world phenomena. ANNs can even create new, synthetic data that augment existing datasets, reducing the need for manual data collection and enabling the simulation of scenarios that are highly valuable in various fields, such as finance, healthcare, and climate modeling. The ability to simulate complex systems and predict the outcomes of different scenarios is a game changer, allowing experts to make more informed decisions, driven by data and insights that were previously inaccessible. This enables them to respond more effectively to a wide range of situations, from responding to natural disasters and predicting market trends to personalized medicine and optimizing business processes. As a result, industries experience increased efficiency, productivity, and competitiveness while also opening new opportunities for innovation and driving growth in sectors that were previously limited by a lack of sufficient data. This challenge has been successfully addressed by the development of ANNs, which have been instrumental in unlocking the potential of complex data by enabling the creation of sophisticated models that can learn from experience, adapt to new situations, and make predictions with the potential to transform industries and revolutionize the way we approach complex problems by providing insights that would otherwise remain hidden.

ANNs have already demonstrated significant potential in various fields and are poised to revolutionize numerous aspects of our lives, from healthcare and education to finance and transportation. They will undoubtedly continue to shape the future of industries and society as a whole, as researchers and developers continue to push the boundaries of what is possible with these powerful tools, driving innovations and breakthroughs that will profoundly impact numerous areas of human life. This progress will fuel unprecedented growth and accelerate the development of transformative technologies that reshape our world and improve the human experience. These include advancements in medical diagnosis and personalized treatment plans, autonomous vehicles that redefine transportation systems, intelligent robots that transform manufacturing processes, and smart cities that optimize energy consumption, waste management, and public services. Ultimately, this will lead to more sustainable, efficient, and livable environments for generations to come, where the seamless integration of AI and the IoT will give rise to entirely new forms of interaction, collaboration, and problem solving. This will enable humans and machines to learn from each other, adapt to changing circumstances, make data-driven decisions that drive innovation, foster economic growth, and create unprecedented opportunities for social mobility, cultural enrichment, and environmental stewardship.

Thus, ANN technology will usher in a new era of unprecedented technological advancements that will forever change the fabric of our society, from the way we communicate to how we work, live, and interact with one another. ANNs form the backbone of this revolution, enabling machines to perceive, process, and respond to complex patterns and data, as well as learn from experience. This makes them capable of simulating human intelligence and decision-making abilities that apply to a wide range of fields, including but not limited to healthcare, finance, education, transportation, and cybersecurity, among others. By facilitating a previously unimagined level of automation, ANNs will propel human civilization forward, with profound implications for the future of work, creativity, innovation, and technological advancements that redefine the boundaries of human potential. This will lead to breakthroughs in fields such as medicine, where novel treatments and personalized healthcare are being developed at an unprecedented rate, transforming the lives of millions of people worldwide and enabling the medical community to address some of the most pressing and complex health challenges of our time, from rare genetic disorders and chronic diseases to infectious diseases and mental health conditions, thus significantly increasing global healthcare solutions.

ANNs are computing systems inspired by the biological neural networks that constitute animal brains. They are a subset of machine learning and form the heart of deep learning algorithms. ANNs consist of interconnected nodes, or "neurons," that are organized in layers. These layers typically include an input layer, one or more hidden layers, and an output layer. Each connection between neurons has an associated weight, which determines the strength of the signal.

Key concepts of ANN:

- **Structure:** The neurons in each layer receive weighted inputs from the neurons in the previous layer. These inputs are summed, and an activation function is applied to the sum to produce the neuron's output. This output is then passed as input to the neurons in the next layer. How deep learning uses ANNs to address complex problems is mentioned [85].
- **Learning:** ANNs learn by adjusting the weights of the connections between neurons. This learning process typically involves feeding the network with a large dataset of examples and using a learning algorithm (such as backpropagation) to iteratively adjust the weights to minimize the error between the network's predicted output and the actual output. Mentions of how neural networks learn patterns.
- **Applications:** ANNs have a wide range of applications, including image recognition, natural language processing, speech recognition, and medical

diagnosis. ANNs provide examples of AI applications in healthcare and gaming.

- **Types of ANNs:** Various types of ANNs are suited for different tasks. Some common types include the following:
  - Perceptron: The most basic form of ANN provides details on the structure of a perception.
  - Multilayer perceptron: Networks with multiple hidden layers can learn complex nonlinear relationships.
  - Convolutional Neural Networks: Specialized for processing images and videos.
  - Recurrent Neural Networks: Designed for sequential data, such as text and time series.

ANNs are powerful tools for solving complex problems, but they also have limitations. One major challenge is "the curse of dimensionality," which refers to the exponential increase in computational complexity as the number of input features increases. Understanding these challenges is crucial for effectively utilizing ANNs. ANNs are a fundamental component of machine learning, inspired by the biological neural networks found in the human brain. These interconnected networks of processing units, known as neurons, can detect complex nonlinear relationships within data, a capability that often surpasses human analytical abilities [85].

Artificial neural networks have found widespread applications in various fields, including computer vision, speech recognition, natural language processing, medical image analysis, and even climate science. These networks possess the remarkable ability to learn from data, adapting and improving their performance over time as new information is introduced. Deep learning architectures, such as deep neural networks, deep belief networks, and convolutional neural networks, have further advanced the capabilities of artificial neural networks, producing results that are often comparable to or even surpass human expert performance [85, 41]. The success of artificial neural networks in recent years can be attributed to several factors, including advancements in learning algorithms, the availability of large datasets, and the utilization of highly parallel hardware, such as graphics processing units (GPUs). Additionally, the plasticity of the brain serves as an inspiration for the adaptability of artificial neural networks, allowing them to learn and evolve on the basis of the available data. As the field of AI continues to rapidly advance, the potential applications of artificial neural networks are likely to expand further, with significant implications for various industries and fields of study.

A particle swarm optimization (PSO)-based ANN is used for frequency regulation of an isolated microgrid that includes electric vehicles (EVs) [120]. A complex power network has utilized an ANN-based PID controller to maintain frequency stability during sudden load variations [83]. An optimal ANN controller has been implemented for interconnected thermal power networks with nonlinearities, such as governor rate constraints (GRCs), for LFC [118]. The LFC of a hybrid microgrid power network has been achieved via the Model Reference Adaptive Controller

(MRAC)-based ANN controller [89]. An ANN-based fuzzy controller has been implemented in a three-area thermal power system with nonlinearities, and the results have been compared with those of a conventional integral controller [14]. An artificial neural network-based proportional integral derivative (PID) controller has been designed for a four-area power grid to regulate frequency oscillations [117]. ANNs and artificial fuzzy systems have been employed to regulate the frequency of the system during unexpected power changes in the grid. The results were compared with those of a conventional PID controller [138]. The genetic algorithm (GA) technique has been used to train the ANN for frequency regulation, and the response of the controller has been compared with that of the classical adaptive control scheme [34]. The above analysis clearly shows that ANNs were used in earlier applications to optimize and regulate the system frequency of isolated and interconnected power grids.

1.1.2 FUZZY SYSTEMS

A fuzzy system for frequency control uses fuzzy logic to regulate and stabilize frequency in power networks. Maintaining a constant frequency in power networks is critical since variations can cause equipment damage, loss of electricity, or even blackouts. Fuzzy logic systems are especially beneficial in this situation because they can address inherent uncertainty and imprecision in monitoring and managing power system parameters. Overview of how a fuzzy system is used for frequency control:

Step 1: Input Variables:
- Frequency deviation (Δf): The discrepancy between the system's real and nominal frequencies.
- Rate of change of frequency: The rate at which the frequency changes, indicating how rapidly the system deviates from the target frequency.
- Other factors: load changes or generation imbalances.

Step 2: Fuzzy Logic Controller (FLC): The fuzzy logic controller interprets the inputs and performs corrective actions on the basis of a set of rules. Usually, the fuzzy rules have the following form:
- A major corrective action is necessary if the frequency deviation is high and if the rate of change is rapid.
- The correction should be minimal if the frequency deviation is modest and if the rate of change is gradual. Instead of using precise numerical values, these rules assist the fuzzy system in understanding the situation in terms of language variables (e.g., enormous, little, quick, slow). It is hence resistant to ambiguities and changes in system settings.

Step 3: Fuzzification: The input values are converted into fuzzy sets.
- Fuzzy sets such as positive big, positive small, negative large, and negative small might be used to categorize frequency deviation.
- There might be fuzzy sets for the rate of change, such as slow, moderate, and quick. With the use of these sets, the system can communicate the input data in linguistic words that control rules more easily.

Step 4: Inference Engine:

- To generate fuzzy output values, the fuzzy inference engine evaluates the fuzzy input sets under the established fuzzy rules. It also assesses the rule basis to determine the proper outputs for control operations (such as load shedding or power modification).

Step 5: Defuzzification:

- Following inference, the fuzzy output must be transformed back into a precise value so that the power system may be modified (for example, by adjusting the load or generation). Defuzzification techniques such as the centroid approach, which determines the output fuzzy set's center of gravity, are used for this purpose.

Step 6: Output:

- The output might be a change in power generation or energy reserve control to return the system to a nominal frequency.

Fuzzy-based model predictive control (MPC) performance against frequency damping is analysed by comparing the resulting MPC in an isolated microgrid (MG) with that of renewable energy sources [72]. A fuzzy logic controller (FLC) has been implemented to regulate the system frequency damping of an isolated MG, which includes solar and photovoltaic (PV) energy sources [115]. In an interconnected thermal power network, the response of the FLC to frequency regulation was analysed by comparing various classical PID controllers with published results, which revealed that the FLC controls the oscillation quickly over another classical controller [128]. The FLC was used for frequency regulation of micro hydropower plants as a secondary controller [71]. FLCs have been used in three areas of thermal power systems for frequency regulation, and a real-time model has also been developed [15]. Microgrid frequency oscillation has been controlled by using FLC, and the performance has been analysed by comparing the results of classical controllers [88]. An adaptive fuzzy approach was applied for a multiarea power system to control and maintain the system frequency, and the supremacy of the proposed approach was confirmed by comparing the results of the classical controller [151].

Fuzzy-based MPC performance in frequency damping is analysed by comparing the results of MPC in an isolated microgrid (MG) with those of renewable energy sources [72]. A fuzzy logic controller (FLC) has been implemented to regulate system frequency damping in an isolated MG that includes solar and PV energy sources [115]. In an interconnected thermal power network, the response of the FLC for frequency regulation was analysed by comparing it with various classical PID controllers from published studies, and it was found that the FLC reduces oscillation more quickly than other classical controllers do [128]. The FLC has been used for frequency regulation in micro hydropower plants as a secondary controller [71]. Additionally, FLCs have been used in three thermal power system areas for frequency regulation, and a real-time model has also been developed [15]. The microgrid frequency oscillation was controlled via FLC, and the performance was analysed via comparison with that of classical controllers [88]. An adaptive fuzzy approach was applied to a multiarea power system to control and maintain system frequency, and the superiority of the proposed approach was confirmed by comparing the results with those of the classical controller [151].

A modern multiarea thermal power system was investigated by FLC for LFC, and the results of various operating conditions were compared [7]. The power quality of a single-area thermal power system was improved by adopting a fuzzy-tuned PI controller as a secondary controller. The role of the FPI controller is to mitigate frequency oscillation [92]. The FLC and PID controllers are involved in the process of system stability during the load disturbance of a power system. A comparison of the results revealed the supremacy of the FLC for LFC [149]. For a wind power plant, the artificial bee colony (ABC) algorithm has been utilized to tune the member function of the FLC for LFC, and the robustness and reliability of the controller have been tested under various working conditions [1]. An interconnected PV solar power plant is investigated via a genetic algorithm-based FLC controller for LFC, and the results of the classical PID, FLC, and GA–PID controllers are presented. The comparison proved that the GA-FLC has a better dynamic response than the other controllers [20]. The Harmony search cuckoo search algorithm-tuned fuzzy PID (FPID) controller controls the frequency oscillation at the time of hasty loading of a two-area nonreheated thermal network. The fuzzy-based PID controller is designed as a secondary controller [44].

A multiobjective function FPID controller was optimized by the artificial bee colony (ABC) algorithm for the LFC of a high-penetration wind farm. For validation, the response of the controller was compared with that of conventional PID and FLC controllers [2]. The frequency of a hydrohydro power plant system was controlled and monitored by an FLC controller. The superiority of the FLC was proven by comparing the results with those of the PI controller [64]. A Type-1 FLC was designed for the LFC of a single-area power system [26]. For grid frequency control, an FLC controller was used in [35]. The fuzzy logic controller was initially implemented independently. Later, the fuzzy system was incorporated into many optimization algorithms. Over time, optimization algorithms have been used to optimize the controller parameters in the fuzzy logic controller. These optimization techniques have also been used to train the member functions.

1.1.3 EVOLUTIONARY COMPUTATION

The crucial role of the evolutionary computation (EC) technique in power system engineering lies in its application to the optimization, control, reliability, and security of the system, ensuring the delivery of high-quality power to consumers under all conditions. The optimization process addresses various critical aspects, including economic dispatch, optimal power flow, unit commitment, integration of renewable energy sources into the existing system, placement of flexible alternating current transmission system (FACTS) devices, and planning for the expansion of transmission networks within power generation systems. The optimal generation schedule for power plants to minimize operating costs while meeting demand is defined through economic dispatch control. The optimal power flow calculation across the grid, aimed at minimizing losses and improving voltage stability, is supported by optimal power flow analysis. Scheduling the on/off status of power plants over a specific period while considering factors such as start-up costs, ramp rates, and maintenance schedules is accomplished through unit commitment. Determining the optimal locations and parameters for flexible AC transmission system devices enhances power system stability and control. The integration of renewable energy sources, such as wind and solar, into the grid, considering their intermittent nature, is optimized via advanced computational techniques. Additionally, identifying the most cost-effective locations and capacities for new transmission lines supports the expansion of transmission networks.

Computational techniques are applied in power systems for controlling various sections, such as automatic generation control, which regulates the output of power plants to maintain system frequency and voltage stability. LFC is implemented to balance the load and generation in real time, ensuring system stability. Similarly, voltage/reactive power control is utilized to manage reactive power sources and maintain voltage levels within acceptable limits. The reliability and security of power generation are achieved through contingency analysis, which evaluates the impact of various disturbances on system stability and identifies critical components. Additionally, preventive maintenance scheduling is employed to optimize maintenance schedules for power system components, minimizing the risk of failure and ensuring system reliability. The key advantages of evolutionary computation techniques in power system engineering applications include flexibility, global optimization, parallel processing, robustness, and the ability to handle complex constraints.

Several power system problems are nonlinear and nonconvex, making it challenging to find the global optimum via traditional methods. Evolutionary computation techniques, with their inherent exploration and exploitation capabilities, can effectively search the solution space to find near-global optima. Power systems involve numerous constraints, such as operational limits, environmental regulations, and grid stability requirements, which EC techniques can efficiently handle. Moreover, EC techniques are generally robust to noise and uncertainties, which are common in real-world power systems. They can also be easily adapted to different problem formulations and objectives. Additionally, many EC algorithms are inherently parallelizable, allowing for significant reductions in computation time for large-scale power systems.

Evolutionary computation techniques have emerged as powerful tools for solving complex optimization and control problems in power system engineering. Their ability to handle intricate constraints, find near-global optima, and adapt to diverse problem formulations makes them invaluable for addressing the challenges faced by modern power grids. As power systems become increasingly complex and integrated with renewable energy sources, the importance of EC techniques will continue to grow. These methods play crucial roles in addressing various optimization problems in power system engineering, including global optimization, handling uncertainty, managing complex constraints, multiobjective optimization, and ensuring robustness and adaptability. Furthermore, EC techniques can be effectively integrated with probabilistic methods to account for uncertainties in renewable energy generation, load demand, and system parameters, making them essential for ensuring the reliable and efficient operation of future power grids.

Power systems are governed by numerous constraints, such as voltage limits, line flow limits, and stability requirements. Evolutionary computation techniques have proven to be highly effective in handling these constraints, often surpassing traditional methods. The adaptability of these methods to changing system conditions and uncertainties makes them particularly suitable for dynamic and real-time applications. Many power system problems are inherently nonlinear and characterized by multiple local optima. EC algorithms, with their robust exploration capabilities, can effectively search the solution space to identify near-global optima. Additionally, power system problems often involve multiple conflicting objectives, such as minimizing costs while maximizing reliability. EC techniques excel at handling multiobjective optimization problems by providing a set of Pareto optimal solutions.

Evolutionary computation techniques have emerged as powerful tools for solving complex optimization problems in power system engineering. Their ability to manage constraints, explore the solution space efficiently, and adapt to changing conditions makes them invaluable for tackling modern power system challenges. These include the integration of renewable energy sources and the development of smart grids. A review of the literature discussing the applications of EC techniques in power system control for ensuring the delivery of a high-quality power supply is presented below.

An improved sine cosine algorithm (SCA)-based adaptive fuzzy assisted proportional-integral-derivative (AFPID) controller was proposed for LFC of an autonomous power generation system, which comprises fuel cells with energy storage units, diesel generators, wind turbines, and solar PV units [114]. Additionally, an elephant herd optimization algorithm (EHOA)-optimized PID controller is employed for the LFC in a single-area nonheat thermal power system. Its performance is evaluated over PID controllers tuned via the genetic algorithm (GA), bacterial foraging optimization algorithm (BFOA), teaching-learning-based optimization (TLBO), and quasioppositional gray wolf optimization algorithm (QOGWOA) in [129].

Additionally, a Kharitonov theorem-based fuzzy logic-designed proportional-integral (PI) controller is developed and presented for the LFC of a single-area power system. The simulation results of the proposed controller are compared with those of a conventional PID controller to measure the supremacy of the proposed method [45]. A hybrid stochastic fractal search with pattern search (hSFS-PS) technique-tuned cascade PI–PD controller is proposed for rectifying the automatic generation control crisis in a multisource single-area power system comprising thermal, hydro, and gas power units, along with plug-in electric vehicles (PEVs). The superiority of the proposed controller is demonstrated by comparing its results with those of optimal control, differential evolution (DE), and TLBO-tuned PI, PID, and cascade PI-PID controllers for the same considered power system [104].

The imperialist competitive algorithm (ICA) optimization technique was employed to design a fractional-order fuzzy PID (FOFPID) controller that was applied to a power system to overcome AGC issues. The performance of the controller was estimated for automatic generation control in multisource, single- and two-area power systems (hydrothermal and gas), with and without the inclusion of a redox flow battery unit. This performance was associated with the response of the I, PI, and PID controllers tuned via the hybrid stochastic fractal search (hSFS), pattern search (PS), differential evolution (DE), and teaching-learning-based optimization (TLBO) techniques in [9]. Additionally, a PID controller is tuned via the gray wolf optimizer algorithm that was developed, and its supremacy is confirmed for AGC in a three-area interconnected power system incorporating a solar thermal power plant (STPP). The simulation results were compared with the responses of the GWO technique-tuned I, PI, and PID controller responses [135].

The adaptive set point modulation (ASPM) method is employed to tune the gain values of the PI controller, which is proposed and analysed for LFC in a two-area power system with an HVDC link. The performance of the proposed controller is compared with that of conventional controllers, such as PI and PID controllers [73]. Furthermore, a modified harmony search algorithm (MHSA) is employed to tune a PID controller, which is evaluated for LFC in an interconnected two-area (hydrothermal) power generating system via the ITAE objective function. The simulation results of the proposed controller are compared with those of the GA-tuned PID controller in [136]. Furthermore, a BAT algorithm-tuned PID controller is proposed and tested for LFC in an interconnected multiarea power generation system. The simulation results of this controller are compared with those of the fuzzy gain scheduling technique-tuned PI controller and conventional controller responses in [131].

The cuckoo search algorithm-tuned PID controller is utilized for LFC in a three-area interconnected power system consisting of two reheat thermal systems and a hydro system. The effectiveness of this proposed technique is compared with that of a genetic algorithm (GA) and a PSO tuned I controller [3]. Additionally, a genetic algorithm fuzzy (GAF) polar fuzzy logic controller is applied for LFC in a 3-area interconnected power system comprising hydro, nuclear, and thermal power generators, and its performance is better than that of fuzzy and conventional PI controllers in [24]. Furthermore, a BAT algorithm-based cascade PD-PID controller is employed for automatic generation control in a multiarea reheat thermal power system. The performance of the P, PI, and PID controllers is evaluated and compared with that of the cascade PD-PID controller in [29].

A beta wavelet neural network (BWNN)-supported proportional integral plus (PI+) controller is analysed for LFC in interconnected thermal power systems. Both systems are equipped with and without fast-acting energy storage devices such as HAE FC and RFB. The system's performance is compared with results obtained via a BWNN-based PID controller [40]. Additionally, a proportional integral derivative with a filter (PIDF) controller is designed and proposed for the automatic generation control of a multiarea thermal power system in a deregulated environment. The simulation results of the proposed controller are compared with those of the fuzzy logic controller [47]. Furthermore, the controllers were tuned via the fruit fly optimization algorithm (FFOA). The controllers, namely, the I, PI, IDD, PID, and PIDD controllers, are developed and assessed for AGC in multiarea, multisource power systems comprising reheated thermal, hydro, and nuclear generation units. The proposed controller's response is compared with the responses of the I, PI, PID, and IDD controllers [94].

The literature review analysis discussed above clearly demonstrates that computational techniques play a crucial role in ensuring high-quality power generation for consumers. These techniques are employed to fine-tune the gain values of secondary controllers in power generation systems to ensure the power quality and stability of the system. The projection optimization methods include the genetic algorithm (GA) [75], PSO [74], gravitational search algorithm (GSA)[125, 74], bacteria foraging optimization (BFO) [6], firefly algorithm (FA) [136, 131], bat algorithm (BA) [115, 20], cuckoo search algorithm (CSA) [3, 40], artificial bee colony (ABC) [49], hybrid genetic–firefly algorithm [38], modified harmony search algorithm [136], hybrid firefly algorithm combined with pattern search [124], stochastic PSO [66], and ant colony optimization (ACO) [75] to [22]. In addition, several combinations of controllers are utilized, including Variable Structure Control [27], Discrete-Mode Control [80], Classical Controllers [101], Fuzzy IDD Controllers [127], FOPID Controllers [142], 2DOF-PID Controllers [126], Robust PID Controllers [133], Fuzzy Logic Controllers [139, 100], Adaptive Controllers [141], PI+ Controllers [40], Optimal Control [21], Dual-Mode Gain Scheduling for PI Controllers [131], Fractional Order Controllers [106], 2DOF Controllers [41], PD-PID Cascade Controllers [28], Integral Controllers [57], Conventional Controllers [58], fuzzy logic controllers [8], and PID Controllers [66] to [63]. Furthermore, these optimization techniques are extensively applied to address various real-time engineering problems.

1.1.4 APPLICATIONS OF COMPUTATIONAL INTELLIGENCE

In the modern world, the utilization of technology has effectively increased across the engineering, medical, transportation, military, and food industries to achieve improved outputs and balance growing demands. Electrical energy plays a fundamental role in addressing these demands and maintaining stability in daily life. The electrical energy division involves three major components: power generation, transmission, and utilization. The integration of computer technology with power systems has directed the construction of smart grids and innovative industrial systems by incorporating small-scale power generation units to balance load demand effectively. Furthermore, modern technology helps in the design and development of complex electrical systems via simulation tools. Additionally, computer technology is instrumental in solving complex optimization problems, ensuring system stability, and selecting suitable controllers for well-organized operations.

Modern computer-based computational intelligence (CI) tools have emerged as powerful supplementary resources for engineers across various fields, including the mechanical, civil, computer, electronics, electrical, pharmaceutical, and biological domains. These tools emulate human brain functions such as information processing, pattern recognition, learning from observations and experiments, memory storage, and retrieval. In power systems, where demand fluctuates due to deregulation and other factors, engineers require effective tools for planning, control, and operation. Recent advancements in CI tools offer robust solutions for optimization, intelligent decision-making, and efficient system operation. They address complex challenges such as interconnected networks, the integration of power electronic components (e.g., TCSC, UPFC, STATCOM), automatic generation control, and intelligent power system management, providing innovative solutions for both consumers and utilities.

Derivative-free optimization techniques such as simulated annealing (SA) and genetic algorithms, along with swarm intelligence methods such as PSO and ant colony optimization (ACO), play crucial roles in the power systems sector, assisting in control, decision-making, and design processes. Neural networks and fuzzy logic systems are particularly effective in managing continuously varying industrial loads and addressing the nonlinear behavior of power systems by offering solutions for load/weather forecasting, system identification, planning, automatic control, and fault detection. Furthermore, swarm intelligence and evolutionary computation techniques contribute significantly to minimizing losses and resolving routing challenges in power distribution systems.

To maintain stability during faulty situations, the rapid isolation of equipment and early fault detection are crucial. Continuous electricity supply and system performance are significantly impacted by faults, which often arise from natural causes beyond human control. Computational intelligence techniques are designed to enhance fault detection processes, improving performance through accurate fault location and system protection. The application of CI methods offers two major advantages: reduced development time compared with traditional approaches and robust system performance, even in the presence of noise or missing data, commonly referred to as uncertainty. Computational intelligence techniques are applied in power systems to address various challenges, including optimal power flow control, voltage regulation, load forecasting, fault detection and diagnosis, generation scheduling, system stability analysis, and predictive maintenance. The most commonly used CI techniques in power systems include artificial neural networks, fuzzy systems, and evolutionary and swarm intelligence methods.

Computational intelligence techniques are implemented for solving LFC and management issues in power systems. Some of the applications are reviewed below. The aggregation of multiple objective functions into a single objective is achieved via a fuzzy system, taking into account environmental and economic objectives along with technical system constraints. Node sensitivity analysis and analytical methods are employed to calculate the objective function for the optimal allocation of distributed generators [91]. The optimal allocation of the DG load power is considered in [25], and it is dependent on the bus voltage. However, the nominal value of the load is constant and does not change depending upon the period. To solve this crisis, the Harmony Search Algorithm (HAS) is utilized with a single objective function.

The control and optimal allocation of a battery energy storage system (BESS) via multiobjective optimization was presented in [107]. PSO techniques are employed to address two key objectives: minimizing power losses and optimizing the total installed capacity of the BESS. Additionally, the optimal charging and discharging strategies for the BESS, along with its optimal allocation, are determined on the basis of an hourly resolution. Total harmonic distortion is addressed as a critical issue, and distributed generation (DG) is analysed as the objective under static load conditions, as discussed in [77]. These challenges are mitigated via the differential evolution method and a genetic algorithm.

The optimal power generation of a virtual power plant (VPP) is determined via a multiobjective optimization method that is based on the PSO technique. The VPP comprises electric vehicle charging stations and distributed generators. In this study, the author utilized an hourly based daily load profile for simulation analysis [26]. Operational costs and pollution emission components are considered two objectives for optimization in [155]. The ant lion optimizer is utilized to address these objectives, considering hourly variations in the daily load. In this optimization process, both the procedure and the issues are implemented within the same programming tool. The minimization of losses and the penetration level of DG are analysed via a single optimization approach to determine the optimal location of the DG, as presented in [143]. This study incorporates load profiles based on daily variations, as well as wind power and PV power generation, into DG modeling. These profiles are evaluated with minute-level resolution across 12 representative days in the proposed research [18].

The reactive power control by the optimal allocation of DGs with single-objective optimization is rectified by applying a genetic algorithm [152]. To find the value of the objective function value, a cosimulation approach was utilized by employing an external power system simulator. In [36], single-objective optimization was performed under constant load conditions to determine the optimal allocation of DG that minimizes power losses. The forward/backwards sweep power flow method is used to evaluate the objective function, and the implementation is developed within the same optimization programming environment. Additionally, various optimization techniques, such as artificial bee colony (ABC), PSO, and harmony search (HAS), are employed to address optimization challenges [99].

The sensitivity analysis method is applied to allocate DG with a single objective function, as described in [97]. The same programming tools are employed for calculating the objective function and performing the optimization process. This study considers the yearly load profile of wind DG production, but it is unclear whether a constant load profile is used for the analysis. The ant line optimization technique was employed to solve the optimal allocation problem of DG with a multiobjective function. In this work, by considering constant loads, the optimal locations for DGs, reactive power control devices, and battery energy storage systems (BESSs) were determined, as discussed in [140]. The mixed-integer conic programming method was used to optimize the allocation of the BESS and DG. In this optimization problem, the hourly resolution of DG production and load demands at a yearly level is considered, as outlined in [13].

In [37], the authors use the PSO technique to solve the optimal allocation of DG. During the process, constant power flow calculations and load considerations are incorporated, with the optimization and coding performed via the same tool. To achieve the optimal allocation of DGs, a hybrid approach that combines two different methods of optimization is used. Specifically, particle swarm optimization–shuffled frog shaping (PSO-SFL) is applied, as described in [145]. In [144], the authors focused on solving the optimal allocation of DGs, electric vehicle (EV) charging stations, and shunt capacitors via a metaheuristic algorithm, specifically the grasshopper optimization algorithm (GOA). The proposed method formulates a single optimization problem with four objectives, which are aggregated into a single function. Additionally, the method consists of two distinct steps, considering the allocation of shunt capacitors and DGs separately, before addressing the allocation of EV charging stations.

The hybrid GA-PSO optimization technique is employed to address the single-objective optimization problem by incorporating three components, including the aggregated objective function discussed in [19]. A genetic algorithm is used to solve the single-objective optimization problem related to the allocation of BESSs and DGs, taking into account daily hourly load shapes, as outlined in [55]. The optimization problem is solved via the Python programming environment in this study. The author used a differential evolution algorithm to address the allocation problem of a single objective function-based DG. In this optimization problem, both the power factor of the DG and its optimal allocation are treated as variables in the optimization process [30]. From the above literature review analysis, CI techniques are effectively applied to power systems for frequency management during sudden load demand situations to maintain power quality and balance power generation with load demand. In addition, the system's reliability and stability are maintained within the limit.

In power generation systems, many issues cannot be effectively solved via conventional techniques, and more data, along with technology, are needed to find appropriate solutions. However, a major drawback is that the solutions generated are not always accurate. To overcome these challenges, more effective solutions for power systems with efficient techniques need to be developed. This computational intelligence (CI) is a better choice for solving the crisis discussed above. The CI offers optimal solutions within a short period and is not limited by its characteristics. Below are some of the key applications of CI in power systems:

1. Power system operation, control, and planning
2. Power plant and network control
3. Power system automation and electricity markets
4. Distribution systems, distributed generation, and forecasting applications

Power System Operation, Control, and Planning: Power system operation includes the coordination of hydrothermal systems, unit commitment, state estimation, economic dispatch, congestion management, maintenance scheduling, and power/load flow analysis, all of which utilize CI to achieve optimal solutions. Power system control involves managing frequency, voltage, tie-line power flow, and stability and assessing dynamic security. Power system planning focuses on ensuring reliability, expanding transmission and generation capacity, and managing reactive power to deliver quality power to all consumers.

Power Plant and Network Control: This category includes the control of fuel cells and thermal power plants. Network control involves the management, sizing, and placement of FACTS devices within the power system.

Power System Automation and Electricity Markets: Power system automation involves fault diagnosis, management, and restoration and ensures network security and reliability. Additionally, it includes bidding strategies, market clearing, and analysis, all of which are essential components of electricity markets in recent days.

Distribution System and Distributed Generation, Forecasting Application: The distribution system applications include planning, operation, and management of demand-side response; smart grid control; and network configuration. The control of solar PV power generation plants, turbine management in power generation utilizing wind, operation of distributed power generation, renewable energy sources, and planning for distribution of power are the key components of DG applications. The AI role is utilized in long-term and short-term forecasting, including solar, wind power, and electricity market prediction applications. Computational intelligence techniques, such as artificial neural networks, fuzzy logic systems, and evolutionary algorithms, have significantly transformed into power system engineering methods to meet recent requirements. These methods provide powerful tools for addressing real-time challenges such as nonlinearity, uncertainty, and large-scale optimization problems in modern power systems. Some of the major applications are listed as follows:

Power System Stability and Control. Neural networks, fuzzy logic controllers, and genetic algorithms are effectively utilized in power system transient stability assessment, voltage stability analysis, generator dynamic control, and LFC, which is based on system load demand to maintain system stability and control.

Load Forecasting. Short-term load forecasting accurately predicts daily or hourly loads, as well as medium- and long-term forecasts, to support strategic planning and investment decisions, which are handled with the support of computational intelligence. It also supports demand-side load management through the use of fuzzy systems.

Power System Optimization. In power systems, computational intelligence helps in economic load dispatch by meeting demand with generation and operational constraints. Optimizing unit commitment improves generation schedules and efficiently solves nonlinear optimal power flow (OPF) problems in generating units.

Renewable Energy Integration. To address the scarcity of fossil fuels and the issue of global warming worldwide, renewable energy has gained increased attention for power generation and its integration into power systems. In this situation, wind and solar forecasting are used to predict power generation from the availability of renewable sources. Grid integration is used to manage the incorporation of renewable energy into the grid. Additionally, microgrid management is utilized to balance distributed energy resources, and all the above issues are effectively solved and handled with computational intelligence methods.

Fault Detection and Diagnosis. With the support of CI methods, fault location identification in power transmission and distribution networks has improved significantly. Protection systems using CI techniques enhance protective measures and self-healing grids. With the support of the CI method, autonomous fault detection and recovery in smart grids are effective.

Smart Grid Technologies. In modern power systems, smart grid technologies play a vital role. In these grids, automation enables self-optimization and automated operation in all load situations. Smart grids facilitate energy storage management by implementing demand response programs to achieve peak load reduction strategies with the support of computational intelligence techniques.

Power Quality Analysis. Detecting and mitigating harmonics in the power supply to maintain power quality is achieved through wavelet transform, neural networks, and hybrid techniques. These methods are also used for real-time voltage sag and swell analysis and power quality monitoring through pattern recognition and intelligent algorithms, and computational intelligence techniques achieve these objectives.

Energy Management Systems. Apart from the generation and utilization of electric power, it is essential to maintain energy balance by equating load demand with generation. Energy management ensures better system stability and high-quality power delivery to consumers. In this context, real-time energy management is applied to optimize energy distribution and utilization. In addition, demand-side management enables better planning for energy consumption, whereas load-shedding strategies during emergencies are addressed through the provision of computational intelligence techniques.

Distributed Generation (DG) Planning and Control. The planning and control of distributed generation have become increasingly important in recent periods to meet load demand and maintain power system stability. Minimizing losses, managing hybrid energy systems, and ensuring voltage regulation through voltage and reactive power management in the optimal sizing and placement of DG units are achieved with the support of computational intelligence (CI) techniques.

Electric Vehicle Integration. In recent years, electric vehicles have played a crucial role in maintaining sustainable energy development by utilizing non fossil fuels. As a result, challenges in power generation, such as charging infrastructure planning for locating charging stations, load impact analysis on power grids, and smart charging algorithms to optimize EV charging schedules, can be effectively addressed and resolved by implementing CI techniques in the power system.

Computational intelligence plays a vital role in power system engineering by addressing complex challenges, including system stability, optimization, fault diagnosis, and the integration of renewable energy sources into power systems. The growing implementation of CI techniques has contributed to the development of more resilient, smarter, and efficient power systems, paving the way for sustainable energy solutions.

2 Computational techniques in power systems

In real life, in addition to all kinds of energy needs, electrical energy plays a vital role in providing comfort for all living things. Electrical energy is converted from other forms of energy sources. For example, in a thermal power plant, heat energy is converted to electrical energy. In a hydropower plant, the kinetic energy available in water is converted into useful electrical energy, and the kinetic energy of wind energy is converted into electrical energy with the help of a wind power plant. In a gas power plant, natural gas is utilized as fuel for generating electric energy. In addition, the generation of electrical power by generating units that maintain the power quality and stability of the system is essential during emergency loading situations. To address these challenges, electricity systems are linked together by tie-line. When a load demand occurs in one of the interconnected systems, power is exchanged via a tie-line to balance the load demand and preserve system stability. In a power system, two kinds of control systems are employed: automatic load frequency control and automatic voltage regulation. In this proposed research book, the LFC scheme is introduced for frequency regulation of power networks during unexpected loading conditions. To date, many studies have been carried out by many researchers on the frequency regulation of single/multiple-source single/multiarea power networks. A literature review on the implementation of LFC in interconnected power systems with different criteria is presented in Chapter 2, Section 2.1.

2.1 LITERATURE REVIEW

Grid-connected power networks are essential for the future to balance power demand and generation. Power demand is always proportional to the development of industries and the modernization of domestics. To meet the power requirements, almost all the major power plants are interconnected. Throughout the world, thermal, hydro, and nuclear sources are mostly used to generate electric power because alternative renewable sources such as solar, wind, tidal, and geothermal sources are lacking. While many challenges are faced in increasing the capacity and size of power plants, the challenges are maintenance of the power plant, pollution, and power quality. Power quality is a major issue faced by all standalone and interconnected power networks during unexpected load variation. To improve the power quality in terms of stability, the load frequency control (LFC) or automatic generation control (AGC) scheme was introduced [79], [84], [32].

LFC or AGC issues in power networks are handled by many researchers with different optimization techniques and controllers. Figure 2.1 shows the structure/flow of the literature survey carried out on the proposed research work. In Section 2.1.1,

DOI: 10.1201/9781003661153-2

a detailed review related to single/multiarea power system LFC is presented. Section 2.1.2 briefly reviews single/multiarea multisource power system LFCs. Section 2.1.3 illustrates the effects of nonlinear components such as the GRC, GDB, and boiler dynamics. A grid-connected power network is essential for the future to balance power demand and generation. Power demand is always proportional to the development of industries and the modernization of domestic sectors. To meet the power requirements, almost all major power plants are interconnected. Throughout the world, thermal, hydro, and nuclear sources are mostly used to generate electric power. Owing to resource limitations, alternative renewable sources such as solar, wind, tidal, and geothermal sources have started to be used in power production. As the capacity and size of power plants increase, many challenges arise, including power plant maintenance, pollution, and power quality. Power quality is one of the major issues faced by both standalone and interconnected power networks during unexpected load variations. To improve power quality in terms of stability, a load frequency control (LFC) or automatic generation control (AGC) scheme has been introduced [79], [84], [32].

LFC or AGC issues in power networks are handled by many researchers via different optimization techniques and controllers. Figure 2.1 shows the structure/flow of the literature survey related to the proposed research work. Section 2.1.1 presents a detailed review of single/multiarea power system LFCs. Section 2.1.2 provides a brief review of single/multiarea, multisource power system LFCs. Section 2.1.3 illustrates the effects of nonlinear components such as the GRC, GDB, and boiler dynamics.

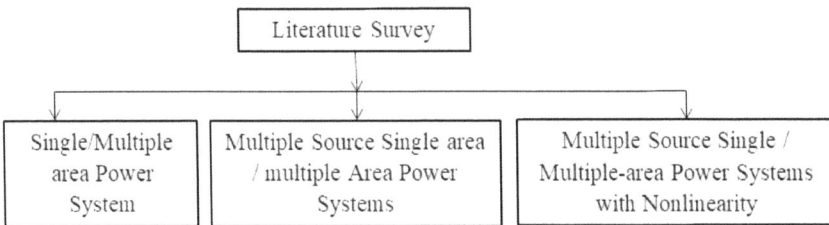

Figure 2.1 Literature survey framework structure

2.1.1 LITERATURE REVIEW ON SINGLE/MULTIPLE AREA POWER SYSTEM LFC

The LFC of single/multisingle single/multiarea multiareas with a single power network is reviewed in detail in this section. An adaptive MPC acts as a secondary controller in [42] to execute the LFC of a 2-area grid connected system with a standalone microgrid (MG), and the results are compared with those of MPC for improvements. A chaos-based firefly algorithm-optimized PID controller was employed in [68] for a two-area hydropower plant (PP) with UPFC for LFC, and the results were compared with those of the GA, PSO, and FFA. The mouth flame optimization (MFO) technique-based fuzzy PID controller was implemented by [23] for LFC in a three-area hydro PP, and the results were compared with those of GA-PID. An ACO-optimized PID controller was implemented in [61] to solve the LFC in a three-area interconnected thermal power network. The symbiotic organism search (SOS) technique-tuned PID controller was applied to a two-area thermal power network for AGC in [53], and the results were compared with those of PSO and DE to prove the improvement of the proposed technique.

The PSO-optimized integral controller is involved in the LFC process of an interconnected thermal power grid with superconducting magnetic energy storage (SMES), and the performance response is equated with the GA and the PSO integral controller without SMES to show the supremacy of the PSO-I with SMES [87]. The author of [81] used a PI controller for a 1-area power network with an MG with a wind form and ESS LFC. The authors of [17] examined a 2-area thermally interconnected electric power network via MFO-PI regulators for AGC, and the performance response was equated with that of the conventional, FA, BFOA, and GWO techniques for identical controllers and systems. The elephant herding optimization (EHO) technique-optimized PID controller was applied by the author in [129] to solve the LFC issue of a reheated thermal power system, and the performance was compared with several of the leading optimization techniques discussed by researchers.

Multiple wind farms are considered with a thermal PP for the ith number of areas of AGC problems by applying the PI as an auxiliary controller [154]. A hydrothermal interconnected power grid is examined via MPC in addition to the effect of the SMES, which is also analysed, and the response is compared via conventional PI [33]. A linear matrix inequality-based PID controller was employed by [137] for the AGC problem of a 4-area reheated thermal power system. Synchronous induction generator-based wind farms are used to control fluctuations in the power grid by varying the output of the wind farms [132]. A firefly algorithm-based PID controller was employed by the author in [69] to control power fluctuations in the power grid, and to prove the supremacy of the FFA–PID, the responses were compared with those of the GA and PSO techniques.

For AGC in a two-area thermal power grid, a tilt integral derivative (TLD) controller was applied [147], and the performance response was compared with the PI and PID controller simulation results. The author of [61] evaluated a three-area thermal power system using a PSO–PID controller to control power fluctuations during unexpected loading periods, and the performance was compared with that of the GA and HC-optimized PID controllers. In [39], the authors utilized the BBO – 3DoFPID controller for power fluctuations in a 2-area thermal-hydro power network. A DEA-tuned FOPID controller was adopted in [146] for the LFC of a thermal–hydro power grid. To control the thermal power system's dynamic reaction, a PSO–PID controller was suggested and demonstrated by [65], and in [63], a thermal–hydroelectric network was investigated by a PI regulator for handling AGC issues in the grid line. A nonreheated, reheated, and double-reheated thermal power network is examined by the PID controller to solve the AGC problem [59]. A two-area thermal PP was examined by a BBO–PID controller, and the response was compared with the PSO and DE performance responses to determine the improvement of the proposed controller [50]. The thermal–hydropower network is evaluated by a 3DoF-PID controller for AGC [111]. Single and double reheated turbines with thermal PPs were examined by a PID controller, which is tuned by the ACO technique for AGC [70]. A four-area thermal power system was investigated with a PSO-PID controller in [119] to solve the AGC problem. Wind from an integrated thermal power plant is examined by the GA–PI controller for AGC [22].

2.1.2 LITERATURE REVIEW ON THE LFC OF MULTIPLE SOURCE SINGLE-AREA/MULTIPLE AREA POWER SYSTEMS

A modified protective controller (MPC) is employed for the LFC issue of the 3-area grid-connected power system [153]. An optimal proportional–integral (PI) controller controls the oscillation of the power supply in a two-area interconnected power network during an emergency [54]. Whale optimization tuned the I+PD cascade regulator utilized by [113] for a nonrenewable source power grid and the PI [82] for an electrical vehicle (EV)-incorporated power network LFC problem. Four areas of deregulated power networks with thermal, gas, diesel and hydropower hydro plants are investigated via a modified virus colony search (MVCS) algorithm-based PID controller for LFC in [43] to compare the results with those of VCS, ABC, and PSO. A single area of hydro, thermal, and gas PPs is subjected to the LFC by the gray wolf algorithm (GWO)-tuned PID, and the response is compared with that of the genetic algorithm (GA) [105].

A hybrid PSO (hPSO) and gravitational search (GS)-optimized PID controller is implemented for two areas of the LFC consisting of (gas, hydro, and thermal PPs) by [76] in addition to a static synchronous series compensator and capacitive energy storage. The bilevel MPC implemented for AGC of a multiarea thermal PP with GRC was integrated with the wind form in [150]. Fuzzy-based differential evaluation-tuned PID was implemented in [5] for the LFC of two-area thermal gas PPs with renewable sources. A whale computation technique-based FPI controller was examined for the LFC of a 2-area hydrothermal and thermal–wind power network in [108], and the performance of the PID controller was compared. A 3-area grid-connected grid power system is investigated via a crew search algorithm-based FOPI controller in [116] for AGC in addition to the HVDC link.

An improved GWO-optimized FPID controller is designed for the LFC of a 2-area source power network, and the I, PI, and PID controller responses are compared [123]. An imperialist competitive algorithm-based FPID controller was employed for the area electric power network AGC problem in [10] to achieve superior performance responses compared with many leading optimization techniques, such as the hFA, GA, and BFOA. The author of [121] used a WAO-tuned 2DOF PID cascade controller as a secondary controller for a three-area electric network to solve the AGC problem, and the results were compared with PID and 2DOF controller simulation results. An improved PSO technique-tuned PID controller was implemented for the interconnected power grid for the LFC issue by [16] in the proposed system, which also considered an HVDC as a parallel source to prove that the IPSO algorithm had the best result compared with the bacterial forging algorithm (BFA) and PSO techniques.

An MFO-based dual-mode controller was examined for an interconnected multi-source (THG–THN) power network for AGC [93]. To prove the dominance of the projected controller, the responses were compared with those of the GA, BFOA, and an MFO technique-tuned PI controller. A genetic-based FPID controller is employed to solve the AGC problem of a 2-area interconnected power network with PVs. The authors [20] compared the results with those of FPID and GA-PID to show that GA-FLPID is best. An ABC-tuned PID controller is employed for the LFC of the interconnected thermal PP, and the results are compared with those of the PSO, EP, GA, and GSA optimization techniques [98]. In [96], a 2-area power network, which includes a thermal–hydrogen gas (PP), was evaluated via FOPID to control the frequency oscillation.

A modified BBO-based PID controller was adopted to solve the AGC of a two-area (thermal–hydro–gas) power system by comparing the response of MBBO with that of the BBO and DE techniques to evaluate the superiority of the proposed controller [51]. A GWO–PID controller was implemented by [52] to adopt LFC for an unequal two-area power network with thermal–hydro and thermal–wind sources. The performance responses were compared with those of a classic controller, ZN, GA, PSO, ABC, and BA. The author of [11] utilized a DISCO PM-based PI controller for the LFC of a hydrothermal power grid. The ACO technique-based PI controller has been implemented in a 2-area thermal–wind interconnected power grid LFC problem, and the response results have been compared with those of the GA and integral controller [4]. The AGC problem of a 2-area power system consisting of hydrothermal methods is achieved by implanting an AC/DC link to the power line [12].

2.1.3   LITERATURE REVIEW ON THE LFC OF MULTIPLE SOURCESINGLE/MULTIPLE-AREA POWER SYSTEMS WITH NONLINEARITY (GRC, GDB, AND BOILER DYNAMICS)

A particle swarm optimization technique-based high-order differential feedback controller (HODFC) was designed in [122] for the LFC of 2 areas: thermal, gas, and hydropower plants (PPs). Population extremal optimization-based PI supremacy was demonstrated for the LFC of a 2-area thermal unit with a GRC, a GDB, and a boiler dynamics in [86]. The slap swarm optimization (SSO) technique-tuned PID controller was employed in [90] to perform the LFC of a 2-area interconnected thermal PP with nonlinearity (GRC, GDB) to compare the supremacy results with those of the GA-PID. Gate-controlled series capacitors (GCSCs) and PI gains are fine-tuned via the optimal fuzzy method in [78] for AGC of a 2-area power grid (thermal, hydro, gas). The performance of FPI is compared with that of the ACO, GA, PSO, and ABC techniques to show the improvement. The teaching-learning-based optimization (TLBO) technique-tuned fractional-order PID has been employed as an auxiliary controller for 2-area hydrothermal power system LFC, and the results were compared with those of TLBO-PID to prove the supremacy of FOPID [48]. A PID controller based on the fruit fly algorithm was constructed for the LFC of a two-area multisource power network, and the simulation results were compared with those of optimal controllers to demonstrate improvement [134]. A modified Jaya optimization algorithm (MJOA)-based integral controller is employed in a 2-area hydrothermal PP LFC with a GRC and a GDB [109], which is also integrated with a wind generator, and the performance response is compared with that of several leading optimization techniques. The ant line technique-tuned fractional order fuzzy-PID (FPID) is designed for the LFC of a 2-area multisource power network consisting of thermal, gas, and hydro PPs [46]. A spider monkey optimization (SMO) technique-based fuzzy logic PID controller was demonstrated in [148] for the LFC of 2-area thermal power systems consisting of a GDB and a GRC. ACO-tuned PID was employed for AGC of 3-area thermal PPs in [67], [103], with 2-area thermal PPs added with nonlinearity components such as the GRC, GDB and boiler dynamics.

The SSA-optimized FOPID controller was implemented for a two-area thermal-hydrogas power network in [95] to solve the LFC problem, and the results were compared with those of PSO, the WOA, and TLBO. To manage the frequency deviation during the rapid loading period, an SOS algorithm-based fuzzy PID controller was deployed in three-area thermal PP with GRC, and the performance was compared via the PSO optimization technique [102]. The PSO–PID controller was implemented in [130] for a 2-area interconnected thermal–wind, thermal–solar power network for the AGC problem. The author compared the results of the I, PI, and PID performance responses to check the supremacy of the proposed controller.

A hybrid DE–GWO technique-optimized FPID controller was implemented for AGC of a 2-area grid-connected power system consisting of (*gas-thermal-hydro unit*) in [31] considering nonlinearity. A biogeography-based optimization (BBO)-tuned 3DoF-PID controller was implemented in [110] and [112] for AGC of 2-area and 3-area power grids, and the 3DoF-PID response was compared with the I, ID, and PID responses to show the improvement of 3DoF-PID. An Oppositional BBO technique-optimized PID controller is employed for the LFC problem of an interconnected power grid, as demonstrated by the author in [52], and the SMES and TCSC are also considered. The ACO technique-optimized PID controller was adopted by the author in [62] to control the frequency deviation in the power grid line for a thermal PP with nonlinearity with different cost functions. A novel anti-wiredup controller is employed in a 2-area thermal power system for LFC by considering the GRC [56]. The cuckoo search optimization technique-tuned PID controller is demonstrated as a three-area nonlinear thermal power system [166]. The DISCO matrix-based PID controller has been used in AGC problems for interconnected power networks (thermal – hydro – nuclear) [94]. The literature reviews carried out in Sections 2.1.1, 2.1.2, and 2.1.3, along with the above reviews, are summarized in Tables 2.1 and 2.2 to identify the research gaps for the proposed research work.

Table 2.1

Literature review of LFC of multi-area multi-sources power network

Ref. No	Thermal PP	Hydro PP	Gas PP	RES	Nonlinearity	I, PI, ID, PID	Another Controller
[153]	✓			✓			✓
[54]	✓	✓	✓	✓			✓
[122]	✓	✓	✓		✓		✓
[113]	✓		✓			✓	
[82]	✓			✓			✓
[43]	✓	✓	✓			✓	
[105]	✓	✓	✓			✓	
[42]	✓						✓
[69]	✓					✓	
[86]	✓				✓	✓	
[76]	✓	✓	✓			✓	
[23]		✓				✓	
[90]	✓				✓	✓	
[150]	✓			✓			✓
[78]	✓	✓	✓		✓	✓	
[5]	✓		✓		✓	✓	
[48]	✓	✓			✓		✓
[134]	✓	✓	✓		✓	✓	
[109]	✓	✓		✓	✓	✓	
[46]	✓	✓	✓		✓		✓
[108]	✓	✓		✓			✓
[116]	✓			✓			✓
[148]	✓				✓	✓	
[61]	✓					✓	
[60]	✓				✓	✓	
[103]	✓				✓	✓	
[53]	✓					✓	
[123]	✓	✓	✓			✓	
[95]	✓	✓	✓				✓
[10]	✓	✓					✓
[121]	✓		✓	✓			✓
[102]	✓				✓		✓
[130]	✓			✓		✓	
[17]	✓	✓		✓		✓	
[87]	✓				✓		
[81]			✓	✓	✓		
[93]	✓	✓	✓			✓	
[16]	✓					✓	
[129]	✓					✓	
[154]	✓			✓		✓	
[33]	✓	✓					✓
[20]	✓		✓	✓			✓

Table 2.2

Literature review of LFC of multi-area multi-sources power network Cont..

Ref. No	Thermal PP	Hydro PP	Gas PP	RES	Nonlinearity	I, PI, ID, PID	Another Controller
[98]	✓				✓	✓	
[137]	✓				✓		✓
[96]	✓	✓	✓			✓	✓
[132]				✓		✓	✓
[68]	✓				✓		
[147]	✓					✓	✓
[67]	✓				✓		
[31]	✓	✓	✓			✓	✓
[110]	✓			✓		✓	✓
[112]	✓					✓	✓
[51]	✓	✓	✓		✓		
[39]	✓	✓			✓		✓
[52]	✓	✓		✓	✓	✓	
[62]	✓				✓	✓	
[56]	✓						✓
[146]	✓	✓					✓
[11]	✓	✓			✓		
[4]	✓			✓	✓		
[12]	✓	✓					✓
[65]	✓				✓		
[63]	✓	✓		✓	✓		
[59]	✓				✓		
[50]	✓				✓	✓	
[94]	✓	✓			✓	✓	
[111]	✓	✓			✓		
[70]	✓				✓		
[119]	✓				✓		
[22]	✓				✓		

On the basis of the above literature survey related to single/multiarea single/multisource multipower system LFC, several works have been carried out that incorporate thermal, hydro, and gas power systems. The analysis was also performed by considering different renewable energy sources with various combinations of controllers, such as integral (I), proportional integral (PI), and proportional integral derivative (PID) controllers, a secondary controller to regulate the stability of the system and the quality of the generated power supply during unexpected loading situations. On the basis of the above literature survey related to single/multiarea, single/multisource power system load frequency control (LFC), several studies have been conducted by incorporating thermal, hydro, and gas power systems. The analysis has also been carried out by considering different renewable energy sources in various combinations with controllers such as integral (I), proportional integral (PI), and proportional integral derivative (PID) controllers as secondary controllers to regulate system stability and ensure the quality of the generated power supply during unexpected loading situations.

3 Mathematical model of power systems

3.1 BLOCK DIAGRAM OF A MULTIPLE-SOURCE, MULTIAREA POWER SYSTEM

Multiple-area multisource power systems with many cases are created and studied in this proposed research. A block diagram of the proposed system is given in Figure 2.1. It consists of multiple areas (two areas) that are all interconnected via a tie-line for sharing generated power during emergency loading situations to maintain the stability and quality of the generated power supply under critical situations. The investigated system comprises two areas, and each area has three different power sources (multiple sources) for providing the required power to the power generating system. In this project, a single-area system is composed of three separate sources: thermal, hydro, and gas. The block arrangement of the examined multiple-area multisource power generating system is shown in Figure 3.1 The detailed components of each power source with various component arrangements are given in Sections 3.1.1, 3.1.2, and 3.1.3 for the thermal source, hydro source, and gas source, respectively.

3.1.1 BLOCK DIAGRAM OF THE THERMAL POWER SOURCE

The component arrangement of the thermal power sources is depicted in the block diagram and effectively shown in Figure 3.2. The components of the thermal power sources are the speed governor, steam turbine, reheater, and thermal unit power-sharing factor. All the above mentioned components are connected to provide power to the system to develop supplies to consumers.

3.1.2 BLOCK DIAGRAM OF THE HYDROPOWER SOURCE

The mechanical hydraulic governor and hydro turbine with power-sharing factors are the components of the hydro source. By utilizing these components, the hydropower source is designed to prove power to the system, and its block diagram is shown in Figure 3.3.

3.1.3 BLOCK DIAGRAM OF THE GAS POWER SYSTEM

The gas turbine speed governor, positioner, fuel system and combustor, and gas turbine unit are among the components of the gas power source unit. Figure 3.4 depicts the gas source's component configuration and block diagram.

DOI: 10.1201/9781003661153-3

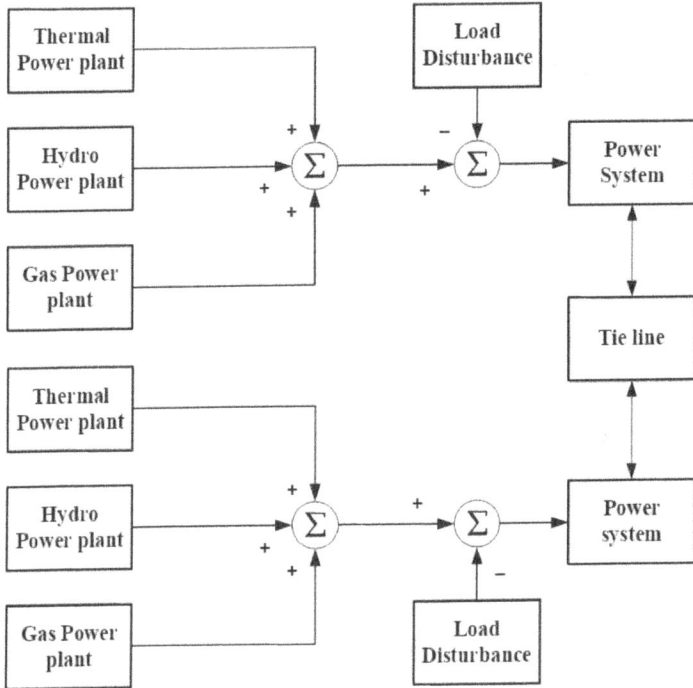

Figure 3.1 Block arrangement of a multiple area multisource power system

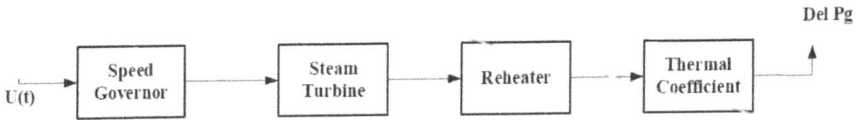

Figure 3.2 Block arrangement of the thermal power source

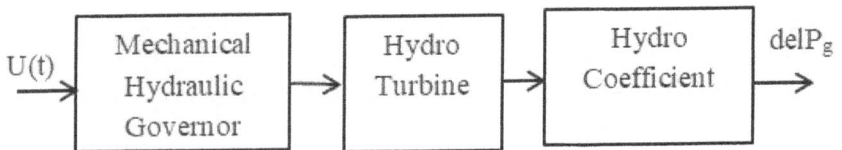

Figure 3.3 Block arrangement of the hydropower source

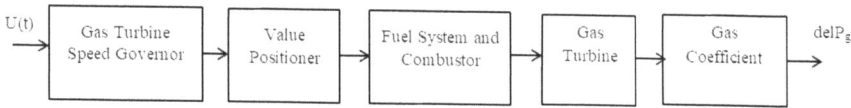

Figure 3.4 Block arrangement of the gas power source

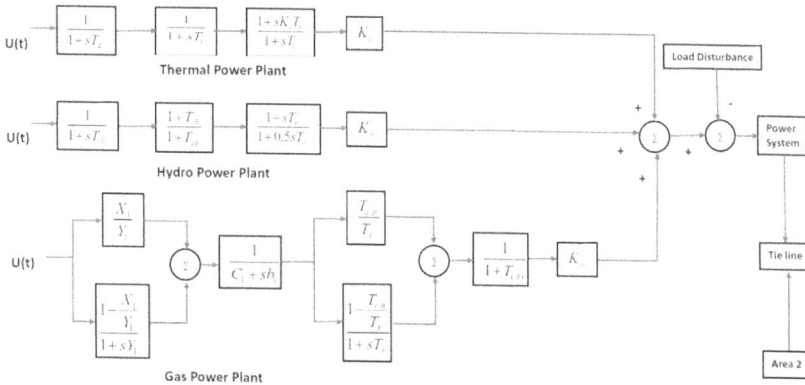

Figure 3.5 Mathematical design of a multiple-source, multiarea power system

3.2 MATHEMATICAL MODELING OF A MULTIPLE-AREA POWER SYSTEM

In this part, the overall arrangement of the mathematical modeling of a multiple-area multisource power generating system is developed in the MATLAB environment, as shown in Figure 3.5. The detailed mathematical modeling of each source is given below in Section 3.2.1.

3.2.1 MODELING OF THERMAL, HYDRO, AND GAS POWER SOURCES

The mathematical modeling of the thermal source, hydro source, and gas source is shown in Figures 3.6, 3.7, and 3.8, respectively. T_{gi} indicates the time constant of the stream turbine governor, the time constant of the steam turbine is represented as Tti, the steam turbine coefficient is represented as K_{ri}, and the time constant of the steam turbine reheater is written as Tri. The mathematical design of the thermal power source is given in Figure 3.6.

The hydro mathematical model of the system is designed by utilizing the following parameters. For example, the transient droop time constant (hydroturbine governor) is denoted as T_{RHi}, Servo time. The constant (hydro turbine governor) is T_{GHi}, the governor reset time (hydro turbine) is T_{Ri}, the penstock (starting time) is T_{wi}, and the arrangement is shown in Figure 3.7.

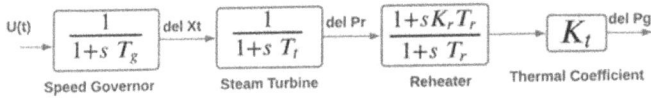

Figure 3.6 Mathematical design of the thermal power source

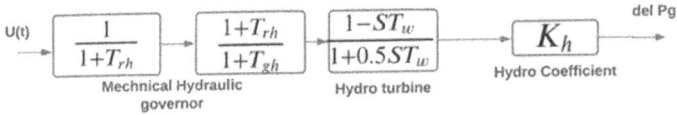

Figure 3.7 Mathematical design of the hydropower source

Similarly, the lead time constant (gas turbine governor) is represented by X_i, the lag time constant (gas turbine governor) is represented by Y_i, the valve positioner constant (gas turbine) is represented by b_i, C_i, the combustion reaction time delay (gas turbine) is represented by T_{CRi}, the fuel time constant (gas turbine) is represented by T_{Fi}, and the discharge volume time constant (gas turbine compressor) is represented by T_{CDi}. The mathematical modeling of a gas source in Figure 3.8 is achieved by using these parameters.

By utilizing the above three mathematical models of thermal, hydro, and gas sources, two areas of interconnected multisource power systems are given in Figure 3.8. Without applying secondary controller action, the open loop behavior of the system is examined and shown in Section 3.3.

3.3 OPEN-LOOP RESPONSE OF A MULTIPLE-AREA POWER SYSTEM

In area 1, one percent step load perturbation (SLP) is used to analyze the open loop behavior of the examined power system. Figures 3.9 to 3.11 depict the system performance (system frequency and tie-line power flow across interconnected systems).

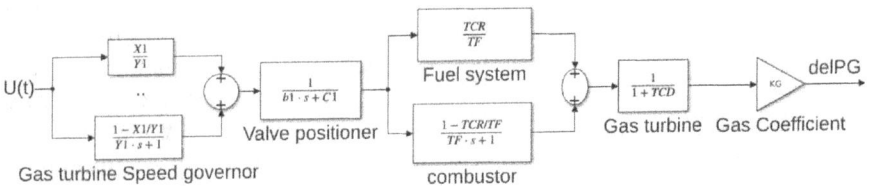

Figure 3.8 Mathematical design of the gas power source

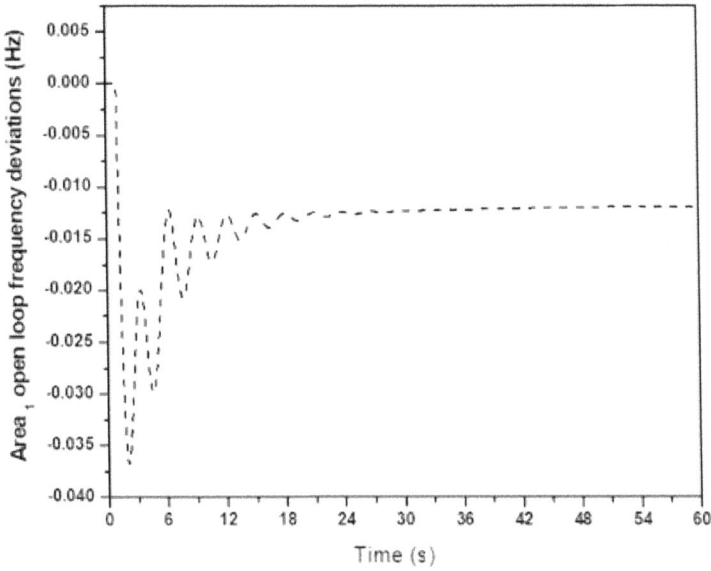

Figure 3.9 Area₁ open loop frequency deviations

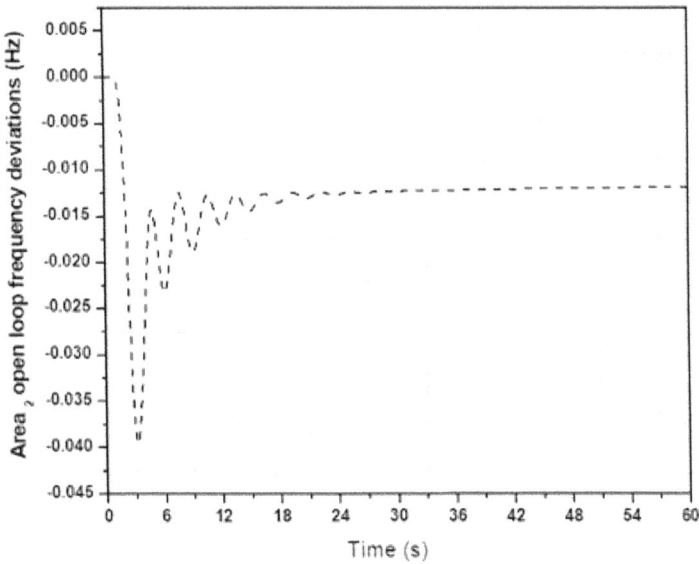

Figure 3.10 Area₂ open loop frequency deviations

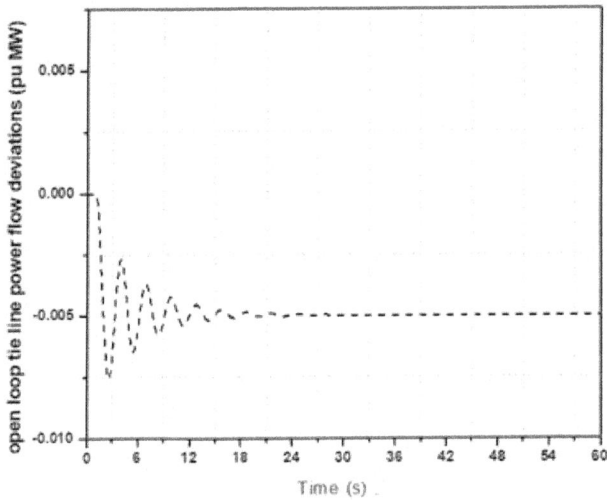

Figure 3.11   Area$_{12}$ open loop tie-line power flow deviations

The numerical values of the open loop response parameters (steady state error, settling time, and peak shoots) of the examined power system are listed in Table 3.1. A bar chart of the settling time is in Figure 3.12

Table 3.1
Numerical values of the open loop response (time domain specification)

Response/Parameter	Settling Time (s)	Peak Over-shoot	Peak Under-shoot	Steady-State Error
Deviations in Area$_1$ system frequency	70	–	-0.037	-0.0116
Deviations in Area$_2$ system frequency	65	–	-0.0398	-0.0118
Deviations in tie-line power flow	40	–	-0.0752	-0.005

The analysis of the system reaction under open loop conditions effectively reveals that when the load demand develops in anyone on the interconnected system, the system frequency and tie-line power flow across the connected power generating system are influenced. This control signal provided by the primary loop available in the system is not sufficient to balance demand with generation by maintaining system stability and the quality of the power supply. To overcome these issues, controllers are implemented in power systems to maintain system stability and supply

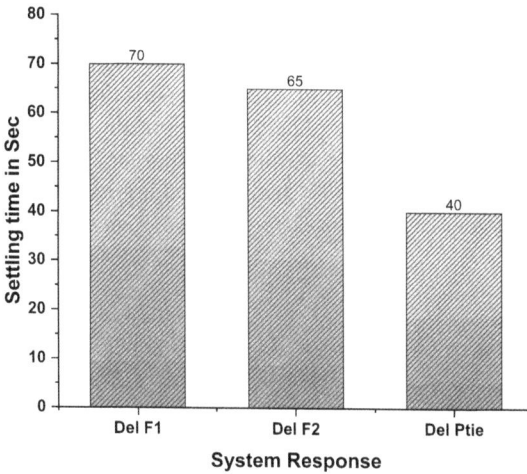

Figure 3.12 Evaluation of settling time in the open loop system

quality. The needs, role of the controller in the power system, and power quality improvement are discussed in Section 3.4.

3.4 NEED OF A CONTROLLER IN A POWER SYSTEM FOR POWER QUALITY IMPROVEMENT

By analyzing the response of the investigated system under the open loop criterion, it is effectively shown that the control action provided by the primary loop is not sufficient during emergencies. Secondary controllers have been incorporated into power systems to address these challenges and preserve system stability and quality. On the basis of the literature review in Section 1.1 under various criteria, a proportional (P) controller, an integral (I) controller, a derivative (D) controller, and, on the basis of these above controller combinations, controllers are applied to regulate the system stability and excellence of the power supply. Additionally, the same combinations of controllers are applied in 2-degree-of-freedom controllers, 3-degree-of-freedom controllers, and fractional-order controllers, and tilt controllers are utilized in interconnected power systems. In addition, several intelligent controllers are also utilized, such as fuzzy logic controllers, and artificial neural networks with a combination of basic controllers are intended and implemented in the power system.

The performance of the controller depends entirely on the suitable parameter values under emergencies. To overcome these issues, several bioinspired algorithms have been applied to tune/optimize the parameter values of the controller. These include the genetic algorithm [142], particle swarm optimization, ant colony optimization, whale optimization [104], gray wolf optimization [105,107], the firefly algorithm [109], the moth flame optimization technique [112], the salp swarm algorithm [113], the differential evolution algorithm [116], the TLBO algorithm [117], the fruit fly algorithm [118], the Jaya algorithm [119], the ant lion algorithm [120], the crow search algorithm [122], the BAT algorithm, the ABC algorithm [143], biogeography-based optimization [154], Oppositional biogeography-based optimization [155], cuckoo search [166], the quasioppositional harmony search algorithm [177], and a novel elephant herd herding optimization [139]. By utilizing the above-described technique, effective controllers are designed to obtain better outputs from the power generating system with stability and a good quality power supply during emergency load demand situations.

3.5 OBJECTIVES

On the basis of the results of the survey, single/multiarea power systems are effectively used to provide high-quality power to consumers. Various power generating sources are included in a single area power system in this connection during times of high demand. The inclusion of more than one power-generating source in a single-area system results in the development of a single-area multisource power system. To obtain a good-quality generated power supply from the system, nonlinear components are included during the design of controller gain values. Additionally, renewable energy sources and energy storage units are included for delivering the required demand power whenever a sudden load occurs in the investigated power system. To achieve the above criteria, different optimization techniques are considered for tuning the PID controller parameters at the time of emergency load demand situations. The primary objectives of this proposed book are as follows:

1. To build an integrated multiarea multisource power-producing system by including a thermal, hydro, and gas unit as a source in a single-area power network.
2. To design a nonlinear multiple area with multiple source-based power systems, nonlinearity effects are considered during the investigation and optimization of controller gain values.
3. Renewable energy sources and energy storage units are proposed for the investigated power systems to improve the quality of the generated supply by keeping the system parameters within the limit.
4. To recommend a suitable computation algorithm/technique for tuning the gain parameters of the controller, the performance is compared to select a suitable tuning method for the design of controller gain values.

4 Different optimization techniques for frequency regulation of multi-source power systems

In this section, the frequency regulation of a multiarea multisource power system is examined by considering the PID controller as a secondary controller. The interconnected power system consists of hydropower generating units, gas power generating units, and thermal power units, which are considered power sources for an area. The same combination of systems is interconnected via tie-line for developing multi (two)-area grid-connected power networks. In this section, the proposed controller gain values are optimized in multiple areas with a multiple-source system by implementing bioinspired algorithms such as the genetic algorithm, firefly algorithm, particle swarm optimization technique, bat algorithm, and ant colony optimization technique by applying a 1% step load disturbance in the analysed power system. The simulation results are compared with each other's technique-based controller performance to select better computation techniques to optimize the gain parameters of the secondary controller. The results efficiently show that the ACO technique-tuned PID controller delivers superior performance over the GA-, PSO-, FA-, BAT-, and algorithm-based PID controllers. The following chapters are organized as follows. The "multisource power system" section provides a power system Simulink model to investigate with a clear transfer function model. The "Objective Function and Controller Design" section provides the details of the cost function applied to find the optimal gain parameters and the effective design details of the controller. The tuned controller gin values with fitness values are reported in the "Tuning of Controller gain value" section, and the results are compared with each other along with a clear effective discussion in the "Results and Discussion" section. Finally, a conclusion concerning the proposed method with different tuning methods is given in the "Conclusion" section.

4.1 MULTISOURCE POWER SYSTEM

The Simulink model of a two-area interconnected multisource power system via MATLAB/Simulink is shown in Figure 4.1, and the proposed system parameters with numerical values are shown in Table 4.1. Each area has a source of THG (thermal, hydro, gas power) in each control area. The areas (control) are interconnected with the AC tie-line, which is useful for exchanging power during emergency load

DOI: 10.1201/9781003661153-4

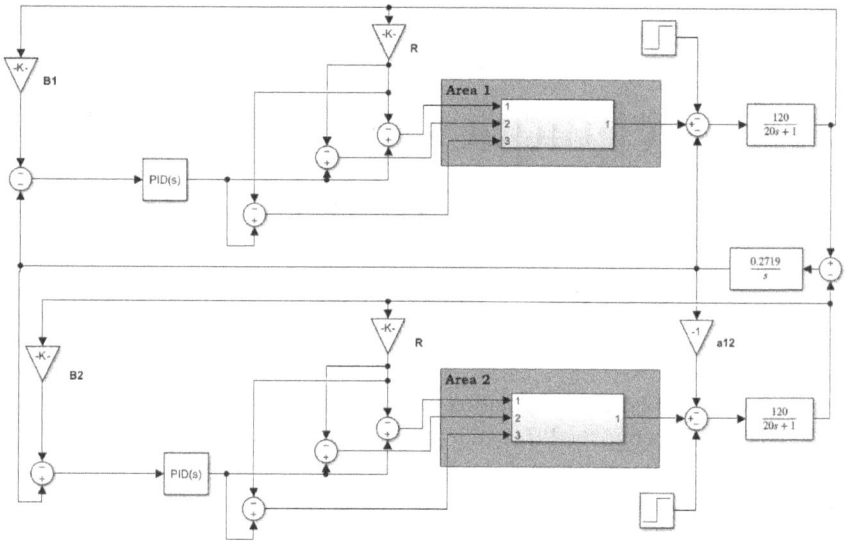

Figure 4.1 Interconnected two area multisource transfer function model block diagram

demand situations when it occurs in an interlinked power system to maintain stability as well as improve the excellence of the generated power supply. The thermal power plant includes the governor, speed turbine, and reheater unit, and the hydropower plant comprises a mechanical hydro governor and hydroturbine units. Similarly, a gas power plant includes a positioner, a fuel system, a combustor, and gas turbine units. The power-sharing factors of the hydro unit, thermal unit, and gas units are denoted as khi, Kti, and kgi, respectively. The mathematical function models of the thermal power, hydropower, and gas power sources are given in Figures 4.2, 4.3, and 4.4 respectively.

Thermal System

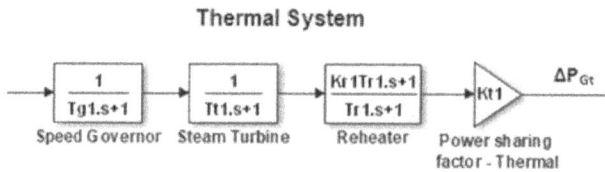

Figure 4.2 Thermal Power plant transfer function model block diagram

The stability of system operation and power mismatch between power generation and total load demand do not occur during nominal loading conditions. However, when load demand develops in any of the interconnected systems, it has an impact on the system's stability as well as the quality of generated electricity. The nominal

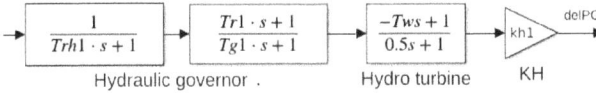

Figure 4.3   Hydro Power plant transfer function model block diagram

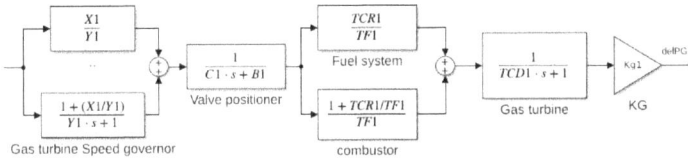

Figure 4.4   Gas Power plant transfer function model block diagram

parameters of the system data with different sources and their values are reported in Table 4.1.

With peak shoots (over and under), the quality of power is impacted in terms of steady-state inaccuracy, and it takes longer to settle during load-demand situations. To address these issues, an appropriate design of the secondary controller is increasingly important for achieving a good quality power supply with stable operation. The proposed controller's design procedure is explained in the next section.

4.2   OBJECTIVE FUNCTION AND CONTROLLER DESIGN

The PID controller is used as a supplementary controller to regulate the examined power system parameters during unanticipated emergency load demand circumstances in this subdivision. The system frequency and tie-line power flow between interlinked power systems are the parameters considered/analysed. The behavior of the controller performance mainly depends on the controller gain values, so the proper selection process of the gain parameters for the controller is most important. The PID controller's transfer function is given in equation 4.1:

$$G_{\text{PID}}(s) = K_p + \frac{K_i}{s} + K_d s \tag{4.1}$$

where,

K_p = Proportional gain.
K_i = Integral gain.
K_d = Derivative gain.
$G(S)$ = PID controller's transfer function.

The genetic algorithm, particle swarm optimization, ant colony optimization (ACO), firefly algorithm (FA), and bat algorithm (BA) optimization approaches are

Table 4.1

Numerical values of the system

System Parameters	Value
Biasing constant b_i	0.425
Regulator constant R_{ti}	2.4 Hz/puMW
Time constant (stream turbine governor) T_{gi}	0.08 s
Time constant (Steam turbine) T_{ti}	0.3 s
Steam turbine coefficient K_{ri}	0.3
Time constant (Steam turbine reheater) T_{ri}	10 s
Transient droop time constant (hydro turbine governor) T_{RHi}	28.75 s
Servo time constant (hydro turbine governor) T_{GHi}	0.2 s
Governor reset time (hydro turbine) T_{Ri}	5 s
Penstock (starting time) T_{wi}	1.0 s
Lead time constant (Gas turbine governor) X_i	0.6 s
Lag time constant (Gas turbine governor) Y_i	1.0 s
Valve positioner constant (Gas turbine) b_i, C_i	0.05, 1.0
Combustion reaction time delay (Gas turbine) T_{CRi}	0.3
Fuel time constant (Gas turbine) T_{Fi}	0.23
Discharge volume time constant (Gas turbine compressor) T_{CDi}	0.2
Generation contribution (thermal)K_{ti}	0.60
Generation contribution (hydro)K_{hi}	0.25
Generation contribution (gas) K_{gi}	0.15
Size ratio of area a_{ij}	1.0
Gain constant of power system K_{Pi}	120.00
Time constant of power system T_{Pi}	20.00 s
Tie-line power coefficient T_{ij}	0.0433 MW

used, and the controller gain values are fine-tuned. The integral time absolute error (ITAE) cost function is used during the tuning of controller gain values. The cost function for ITAE is stated in equation 4.2.

$$J_{\text{ITAE}} = \int_0^T t \cdot |\Delta f_1(t) + \Delta f_2(t) + \Delta P_{\text{tie}}(t)| \, dt \qquad (4.2)$$

where Δf_1 and Δf_2, ΔP_{tie} represent deviations in the frequency in area - 1, area - 2, and tie-line power flow between interconnected systems, respectively; J indicates the performance index; and t represents the time given to simulate the model. The details of the optimization algorithm are given in section 4.3 along with the structure of the controller and the cost function for tuning the controller.

4.3   OPTIMIZED CONTROLLER GAIN PARAMETERS

The conventional PID controller is not sufficient to control and monitor the frequency oscillation of the system during emergency loading conditions. To improve the controller efficiency, the controller parameters need to be optimized. To optimize the controller parameters, different optimization techniques have been used by many researchers. The work of optimizing the proposed controller via the ant colony optimization technique is discussed. The details of the optimization technique and its procedure during the optimization process are clearly explained in Chapter 4, Section 4.3.1. To demonstrate the supremacy of the proposed optimization technique (ACO), different optimization techniques are utilized to tune the proposed controller.

4.3.1   ANT COLONY OPTIMIZATION (ACO) TECHNIQUE

ACO was proposed in 1992 by Marco Dorigo in his Ph.D. research thesis. First, the algorithm was developed to find the optimal way in a graph depending on the behavior of ants for searching paths between food sources and colonies of ants. In general, naturally, ants initially randomly wander around their colony to find the source. During the return trip, the ant lays the pheromone chemical in this track. By reducing the random search for a food source, it is more useful and supportive for other ants to find such paths. Over time, the strength of the chemical starts to decrease because of its evaporation nature. Automatically, the longest path and lesser pheromone layer path are eliminated, and the shortest path with a good-quality food source path has more pheromones than the other paths do. Finally, the shortest path with a good-quality food source has more pheromones. For these reasons, all the ants are attracted and follow the single shortest path. Finally, this natural behavior of real ants has attracted researchers to develop ant colony optimization techniques for solving complex problems. In this book, the ACO computation algorithm is proposed to optimize the gain parameter of the PID controller for the LFC of the grid-connected power network. The flow of the ACO technique during the optimization of the PID controller is as follows:

- Phase 1: Start the process.
- Phase 2: Parameter initialization of the ACO technique.
- Phase 3: Run the designed Simulink model
- Phase 4: Cost function estimation
- Phase 5: Updation of pheromone chemical and probability
- Phase 6: Calculate the gain values of Kp, Ki, and Kd
- Phase 7: Check the iteration level (maximum level or not)
- Phase 8: If it is Yes = stop the process, if it is NO = Go to phase 3 and repeat the same
- Phase 9: Stop the process

By utilizing the optimization techniques (GA, PSO, FA, BAT, and ACO), the controller gain values are tuned, and the gain values are listed in Table 4.2 along

Table 4.2

Gain values of the GA-tuned, PSO-tuned, FA-tuned, BA-tuned and ACO technique-tuned PID controller gain values with fitness values

Technique/ Gain	Kp1	Ki1	Kd1	Kp2	Ki2	Kd2	Fitness Value J
GA	0.9462	0.9064	0.4503	0.6866	0.5825	0.7194	0.5402
PSO	0.7514	0.9999	0.4457	0.9797	0.5358	0.9089	0.5110
FA	0.9999	0.9999	0.6156	0.9093	0.6578	0.8435	0.4918
BAT	0.9997	0.9996	0.4940	0.9936	0.5430	0.6922	0.4887
ACO	0.9999	0.9999	0.5240	0.9999	0.9999	0.9999	0.4831

with the fitness values. Compared with the GA, PSO, FA, and BA technique-tuned controller fitness values, it is obvious from the numerical data in Table 4.2 and the bar chart assessment in Figure 4.5 that the ACO technique-tuned PID controller gives the lowest fitness value.

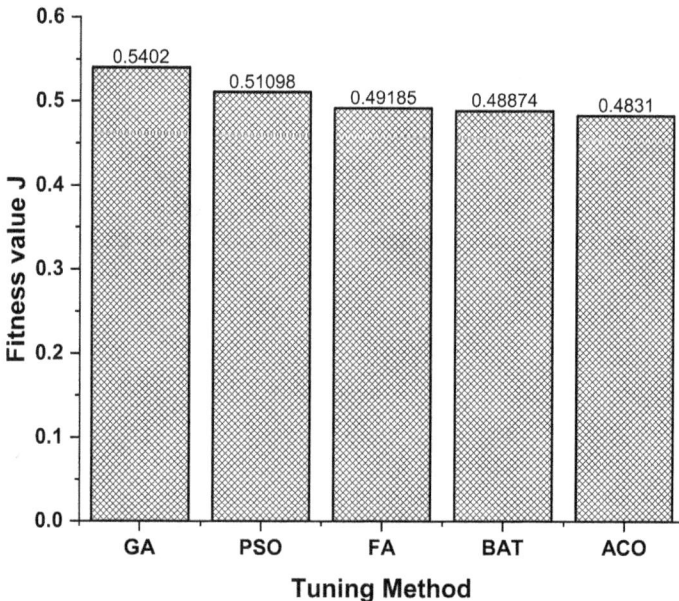

Figure 4.5   Bar chart evaluation of the fitness value J

The performance analysis of the PID controller gain values of different optimization techniques is given in Section 4, Section 4.4. During the investigation of the optimization technique, 1% step load perturbation (SLP) is applied as a load demand in area 1.

4.4 RESULTS AND DISCUSSION

The Simulink model of an examined multiple area with a multisource power system is designed in the MATLAB platform. The behavior of a system depends entirely on the controller's behavior during emergency load demand situations by setting optimal gain values on the basis of the load demand. Different optimization strategies are used to optimize the gain values in this process. To adjust the controller gain values, the GA, PSO, FA, BA, and ACO tuning procedures are suggested in this chapter. The tuned controller gain values with different technique behaviors are shown to select a better optimization technique given in Chapter 4, Section 4.4.1.

4.4.1 PERFORMANCE ANALYSIS

In this part, the performance of different optimization strategies for tuned PID controllers (GA, PSO, FA, BAT, and ACO) is examined to choose the best optimization technique for tuning controller gain values in a power system under sudden load demand conditions.

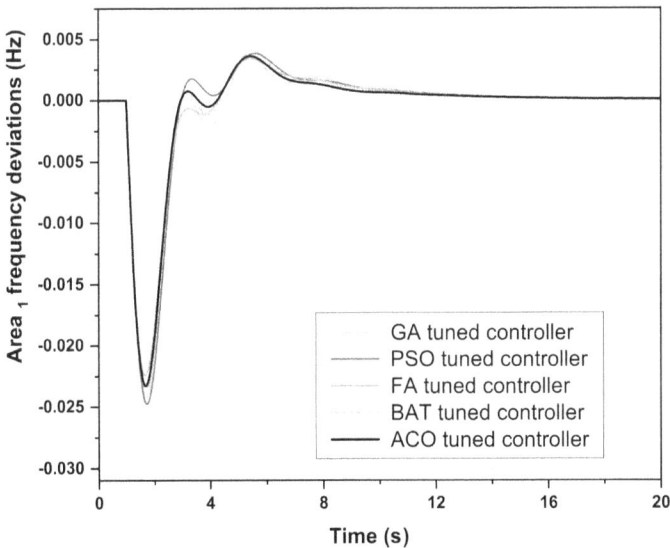

Figure 4.6 Assessment of frequency deviations in area 1

The estimation of the performance of different optimization-tuned controllers is shown in Figures 4.6 to 4.10. In that evaluation of the system, the frequency is given in Figures 4.6 and 4.7, the power flow in the tie-line evaluation is demonstrated in

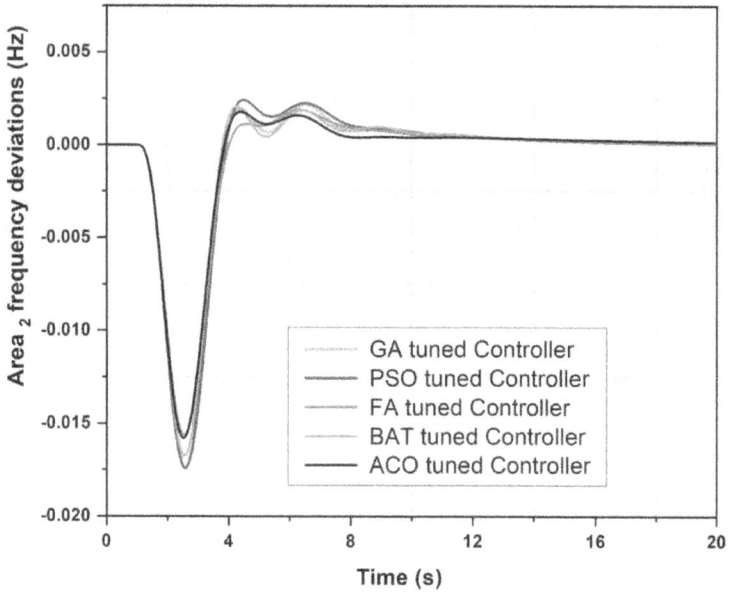

Figure 4.7   Assessment of frequency deviations in area 2

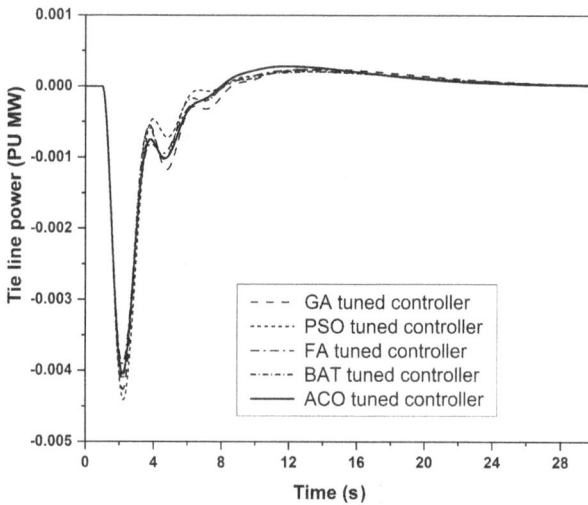

Figure 4.8   Assessment of power deviations in the tie-line between areas 1 and 2

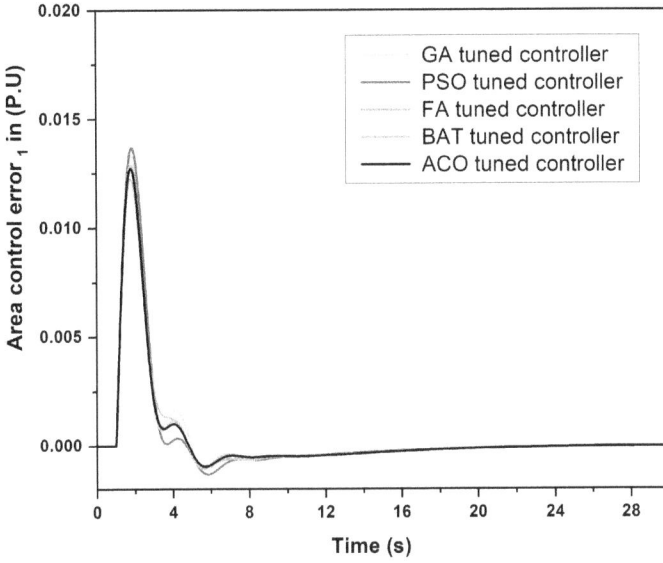

Figure 4.9   Assessment of area 1, area control error

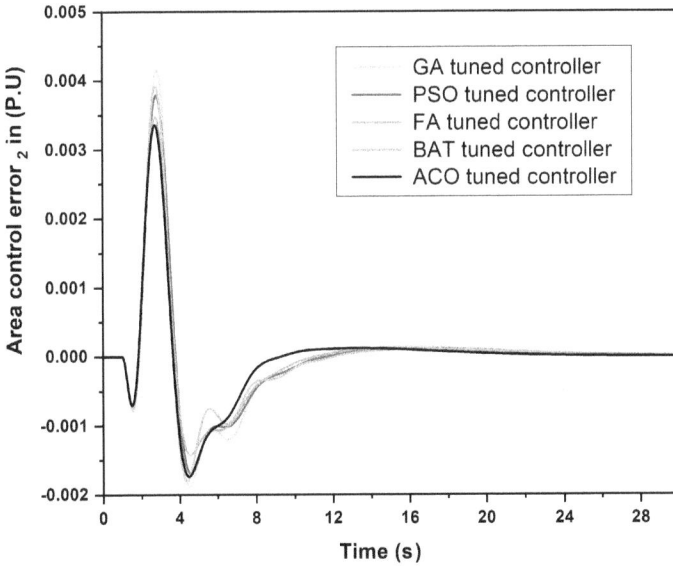

Figure 4.10   Assessment of area 2, area control error

Figure 4.8, and finally, the area control errors (ACEs) of areas 1 and 2 are shown in Figures 4.9 and 4.10, respectively. An analysis of the system response clearly reveals that, compared with other technique-tuned controller parameters, the ACO-PID controller provides better results in terms of settling time and peak shoots during emergency loading conditions. The ACO technique-tuned controller takes 17 s and 25 s to settle frequency deviations in areas 1 and 2 and 29 s to settle deviations in tie-line power (del Ptie) between connected areas. Similarly, the area control error in areas 1 and 2 takes 31 s and 25 s, respectively, to settle the response under stable conditions. The time-domain specification parameters are the settling time (Ts) and peak shoots (over (Os) and under (Us)) of each optimization technique, and comparisons are analysed in Section 4, Section 4.4.2.

4.4.2   TIME DOMAIN PARAMETERS OF THE SYSTEM WITH DIFFERENT OPTIMIZATION TECHNIQUES TUNED WITH THE PID CONTROLLER

The mathematical values of the time domain specification parameters are listed in Tables 4.3, 4.4 and 4.5 for the system frequency in both areas, del Ptie, and ACE in areas 1 and 2, respectively.

Table 4.3

Evaluation of the time domain parameters with respect to the system frequency

Optimization Technique	ΔF_1			ΔF_2		
	TS (s)	OS (Hz)	US (Hz)	TS (s)	OS (Hz)	US (Hz)
GA	20	4×10^{-3}	0.024	33	2.2×10^{-3}	0.0175
PSO	18	3.9×10^{-3}	0.024	30	2.4×10^{-3}	0.0174
FA	19.5	3.5×10^{-3}	0.022	40	1.9×10^{-3}	0.0156
BAT	19	3.8×10^{-3}	0.024	30	2×10^{-3}	0.0170
ACO	17	3.6×10^{-3}	0.023	27	1.8×10^{-3}	0.0155

Table 4.4

Evaluation of time domain parameters for ΔP_{tie}

Optimization Technique	TS (s)	OS (pu MW)	US (pu MW)
GA	35	2.5×10^{-4}	4.3×10^{-3}
PSO	35	2×10^{-4}	4.4×10^{-3}
FA	39	2.2×10^{-4}	3.9×10^{-3}
BAT	36	2×10^{-4}	4.1×10^{-3}
ACO	29	3×10^{-4}	4×10^{-3}

An analysis of the mathematical values in the above table and evaluation comparisons in the above section prove that the ACO method tuned controller settles the

Table 4.5

Evaluation of the time domain parameters in terms of the area control error

Optimization Technique	ACE 1			ACE 2		
	TS (s)	OS (pu)	US (pu)	TS (s)	OS (pu)	US (pu)
GA	33	0.0133	1.2×10^{-3}	31	4.2×10^{-3}	1.9×10^{-3}
PSO	31	0.0137	1.3×10^{-3}	30.5	3.8×10^{-3}	1.7×10^{-3}
FA	32	0.0125	1×10^{-3}	29	3.5×10^{-3}	1.4×10^{-3}
BAT	32	0.0129	1.1×10^{-3}	31	3.9×10^{-3}	1.8×10^{-3}
ACO	31	0.0127	1×10^{-3}	25	3.4×10^{-3}	1.7×10^{-3}

response in the shorter period (f1 $=17$ s, f2 $= 27$ s, tie-line $= 29$ s, ACE1 $= 31$ s, and ACE2 $= 25$ s). This demonstrates that the ACO PID controller takes less time to respond than the GA, PSO, FA, and BAT method-tuned controllers do. To show the supremacy of the ACO technique, a PID controller bar chart is plotted for different technique-tuned controller settling times, and it is depicted in Figures 4.11 to 4.15.

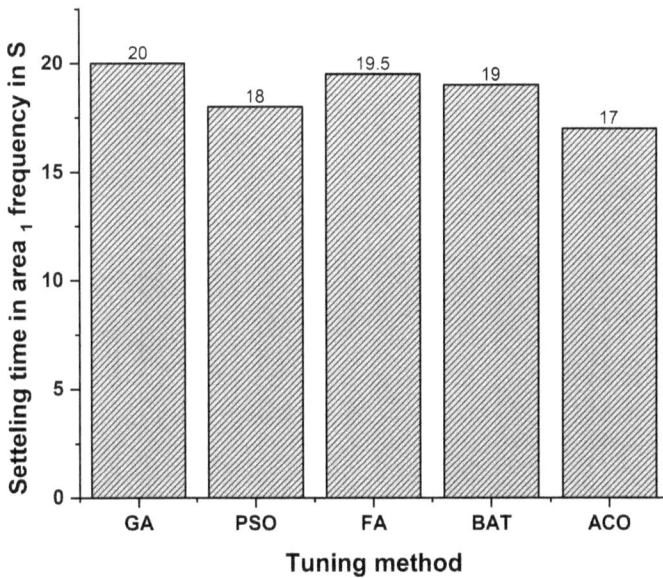

Figure 4.11 Bar chart evaluation of the settling time of the system frequency in area 1

The above bar chart comparison of settling time in terms of system frequency, tie-line power flow variations between systems, and area control error clearly shows that the ACO–PID controller provides a more regulated response during sudden load scenarios in an interconnected power system. In Section 4.4.3, the performance enhancement of the ACO - tuned PID controller is discussed.

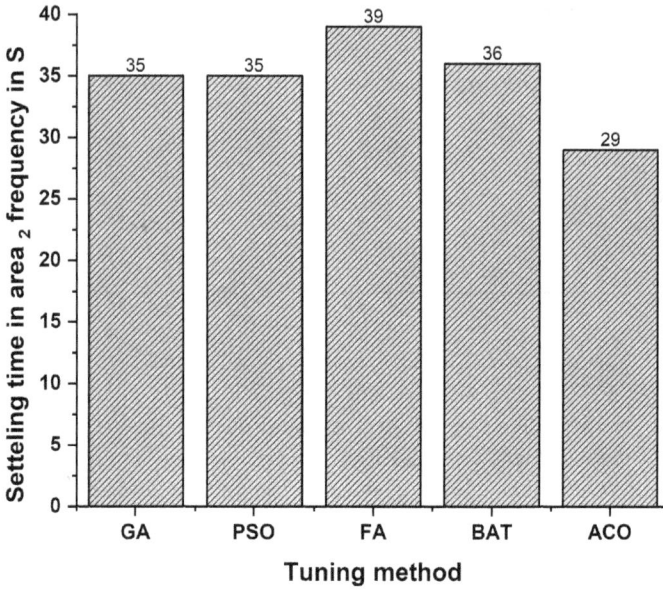

Figure 4.12   Bar chart evaluation of the settling time of the system frequency in area 2

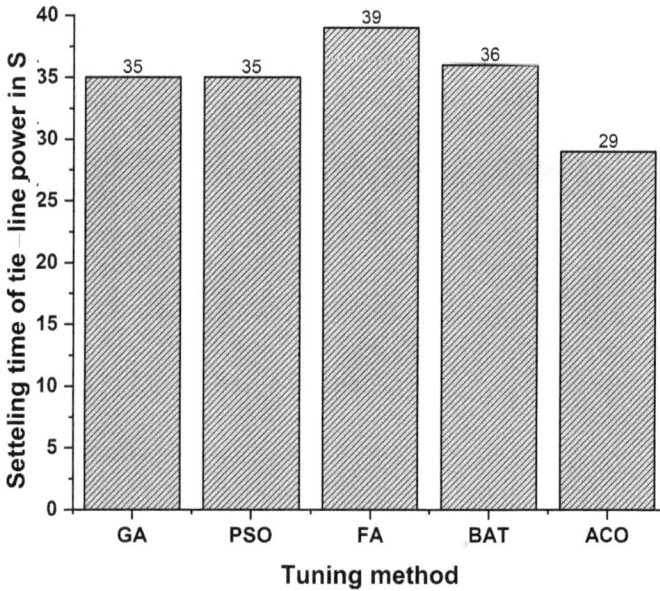

Figure 4.13   Bar chart evaluation of the tie-line power settling time

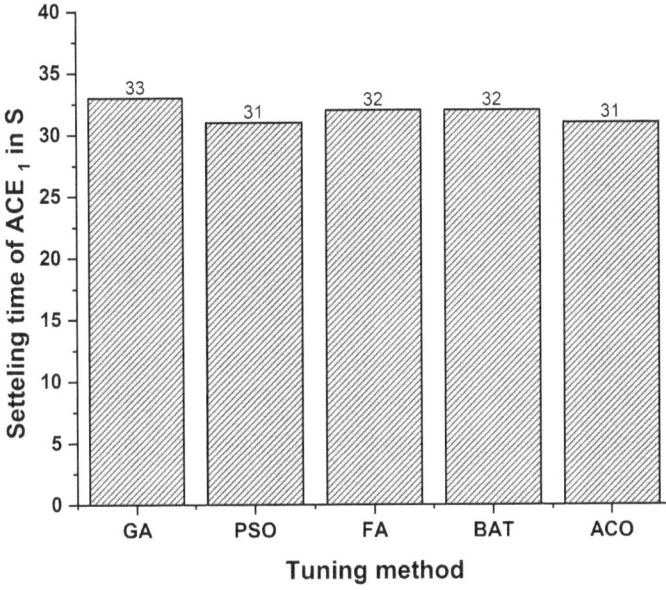

Figure 4.14   Evaluation of the settling time via a bar chart of the area control error in area 1

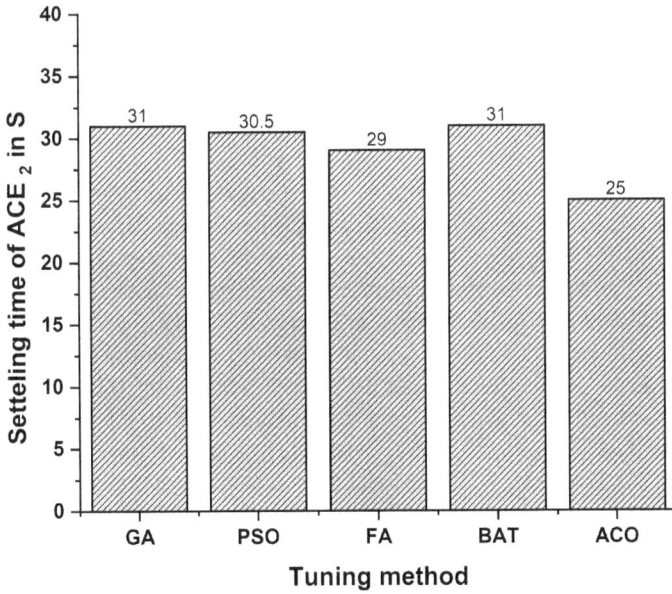

Figure 4.15   Evaluation of the settling time via a bar chart in ACE at area 2

4.4.3 ACO-TUNED PID CONTROLLER PERFORMANCE IMPROVEMENT OVER THE GA, PSO, FA, AND BA TECHNIQUE-TUNED PID CONTROLLERS IN SETTLING TIME

The performance increase of the ACO approach-tuned controller over other techniques, such as the genetic algorithm, particle swarm optimization technique, firefly algorithm, and bat algorithm, are evaluated in this section, with numerical values given in the table below and figures.

Table 4.6

Evaluation of the time-domain parameters with respect to the system frequency

Technique	ACO–PID controller improvement (%) in area 1 system frequency	ACO–PID controller improvement (%) in area 2 system frequency
ACO over GA	15	18
ACO over PSO	6	10
ACO over FA	13	33
ACO over BA	11	10

Concerning the performance improvement analysis in Table 4.6 above and the bar chart evolution in Figures 4.16 and 4.17. It is evident that the ACO-tuned controller provides a better-controlled response in terms of minimal settling, which takes more time than other behaviors do.

On the basis of the performance enhancement analysis of ACO-tuned controller action, the numerical values are listed in Table 4.7, and the bar chart evolution is shown in Figure 4.18. An ACO-tuned controller offers a better controlled response in terms of minimal settling, taking more time than other controllers do.

To prove the superiority of the ACO controller, numerical values of improvement are tabulated in Table 4.8, and the bar chart evolution in Figures 4.6 and 4.7 clearly shows that the ACO-tuned controller delivers a healthier controlled response. Using the evaluation response analysis in Figures 4.6 to 4.10, the numerical values of the time-domain specification parameters and fitness values for each optimization technique can be determined, and the bar chart evaluation graphs of the settling time in Figures 4.11 to 4.15 prove that the ACO technique tuned the proportional integral derivative controller to yield a superior response (minimal settling time and peak shoots) over the GA, PSO, FA, and BAT technique-optimized controller performance in the emergency load demand situation in multiple interlinked areas with a multisource power system. The percentage of improvement in Ts of the proposed method is shown in bar chart in Figure 4.19 to 4.20.

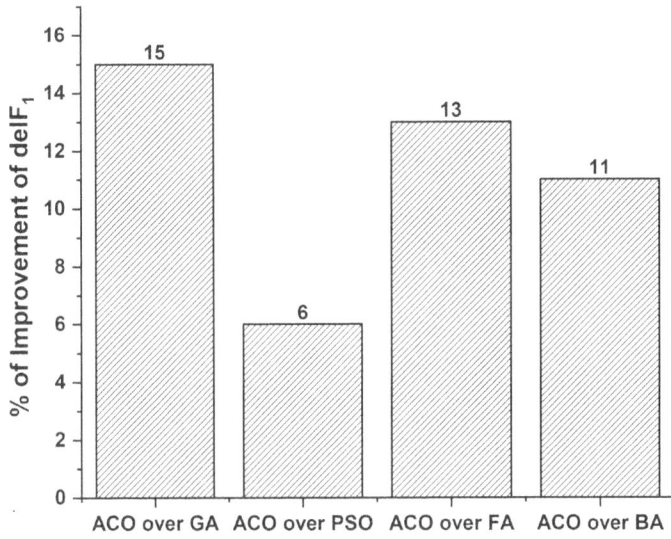

Figure 4.16   Evaluation of the settling time via a bar chart of the system frequency in area 1

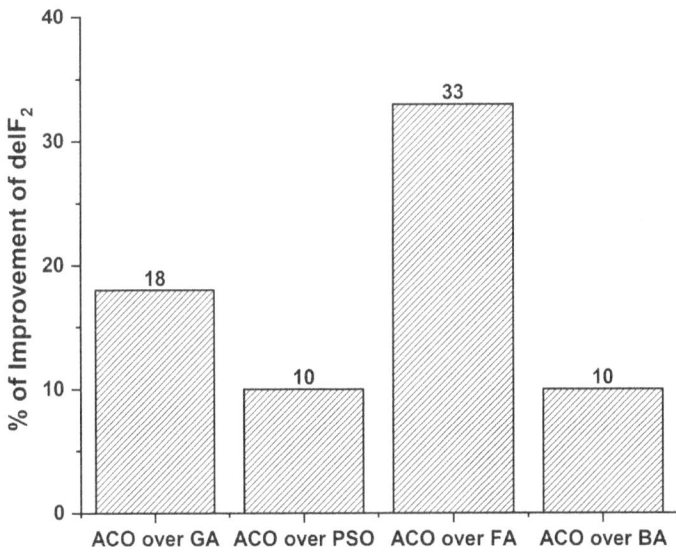

Figure 4.17   Evaluation of the settling time via a bar chart of the system frequency in area 2

Table 4.7

Evaluation of time domain parameters in tie-line power flow

Technique	ACO–PID Improvement in Tie-Line Power Flow (%)
ACO over GA	17
ACO over PSO	17
ACO over FA	26
ACO over BA	19

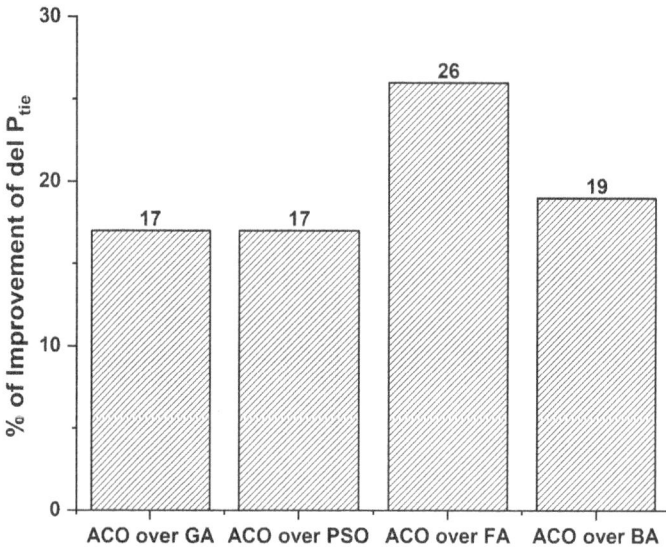

Figure 4.18 Evaluation of the settling time via a bar chart of the tie-line power flow

Table 4.8

Evaluation of the time domain parameters in terms of the area control error

Technique	ACO–PID controller improvement (%) in area 1 ACE	ACO–PID controller improvement (%) in area 2 ACE
ACO over GA	6	19
ACO over PSO	0	18
ACO over FA	3	13
ACO over BA	3	19

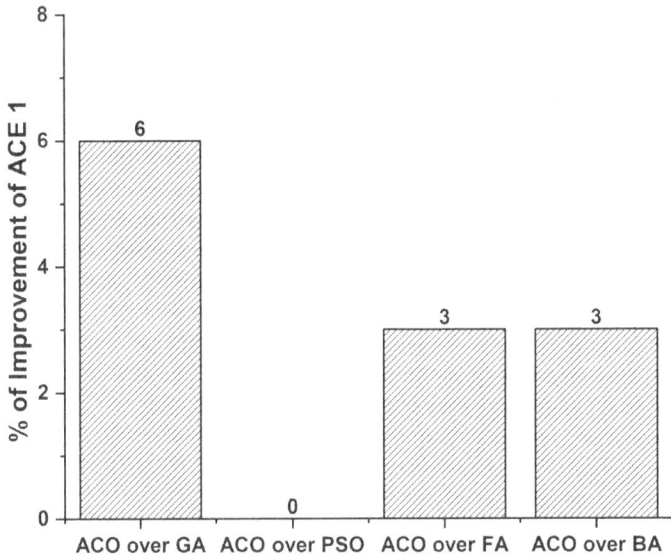

Figure 4.19   Evaluation of the settling time via a bar chart of the area control error in area 1

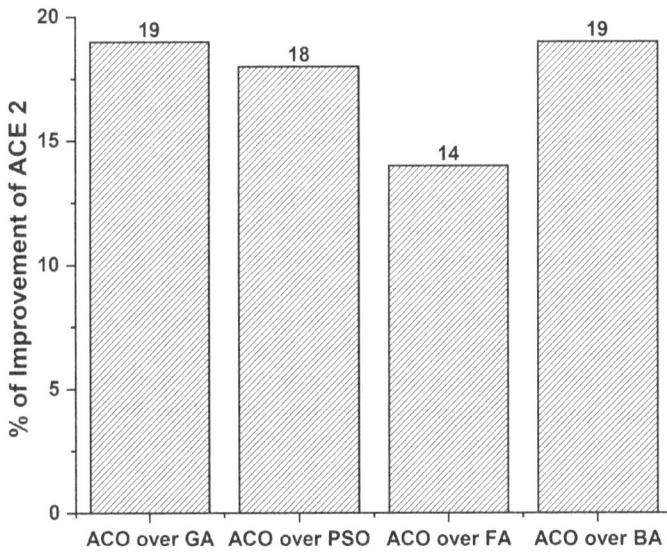

Figure 4.20   Evaluation of the settling time via a bar chart of the area control error in area 2

4.5 CONCLUSION

In the LFC control of two-area interconnected multisource power systems, this chapter investigates the use of various optimization strategies for the construction of secondary controller gain values. Two single areas, each with numerous power-producing units, such as thermal, hydro, and gas units, form a linked two-area power system. In this power system, a PID controller is employed as a secondary controller to investigate LFC systems, and controller gain values are optimized via genetic algorithms, particle swarm optimization techniques, firefly algorithms, bat algorithms, and ant colony optimization techniques. In Region 1 of the analysed system, the controller gain parameters are optimized under 1% SLP.

System performances are compared, and time-domain specification parameters are studied to demonstrate and evaluate the superiority of optimization techniques. The ACO technique-adjusted PID controller requires less time to settle the response (f1 = 17 s, f2 = 27 s, tie-line = 29 s, ACE1 = 31 s, ACE2 = 25 s) with small peak shoots (over and under). The ACO–PID controller clearly has fewer oscillations and a faster settling time than the GA, PSO, FA, and BAT method-tuned controllers during power-generating unit emergencies.

5 Impact of renewable energy sources in multiarea multi source power systems

In this chapter, the frequency regulation and performance of multiarea and multi-source power systems incorporating a renewable energy source (wind) are described in detail. Section 5.1 describes the mathematical model of the multiarea multisource power system with a wind source. Section 5.2 presents the simulation results and discussion, including the performance, time domain-specific parameters, and impact analysis of the proposed system using the ACO–PID controller. The conclusion of the chapter is given in Section 5.3.

5.1 MATHEMATICAL MODEL OF A WIND SOURCE-INCORPORATED MULTIAREA MULTISOURCE POWER SYSTEM

Many renewable energy sources, including wind and solar facilities, have been incorporated into the system to satisfy electricity demands. Wind energy is now one of the fastest-growing renewable energy sources in the world because of its clean and nonpolluting nature. Wind energy is a renewable energy source that converts the kinetic energy of wind into mechanical power or electricity. This procedure is often achieved via wind turbines, which are machines with large blades that revolve as the wind blows.

Benefits of Wind Energy:

- **Sustainable resource:** Wind is an endless resource, making it a sustainable energy source.
- **Low operating expenditures:** After the initial installation, operating and maintenance expenditures are minimal.
- **Environmental impact:** Wind energy emits no greenhouse gases during operation, hence contributing to a lower total carbon footprint.
- **Economic benefits:** Wind energy has the potential to provide jobs in wind turbine production, installation, and maintenance.

The power system's issues have grown as a result of the following factors: (a) increasing integration of renewable sources, (b) deployment of novel technologies such as microgrids, and (c) reliance on insecure communications infrastructure. These

DOI: 10.1201/9781003661153-5

variables have a direct influence on the performance, dependability, and stability of electrical power systems [173]. In this chapter, the wind power plant is integrated with a conventional source power system, which is explained in Chapter 2. The wind power lifts and turns the blades. The whirling blades drive a shaft within the nacelle that feeds into a gearbox. The gearbox increases the generator to convert rotational energy into electrical energy via magnetic fields; therefore, the rotational speed is adjusted accordingly [174]. As wind power becomes more incorporated into power networks, the uncertainty associated with active power generation increases, resulting in greater frequency variations. To solve these issues, future LFC approaches for power systems need to be more durable and efficient. Wind speed fluctuations can influence the power output of WTs, which can affect the grid's balance of power generation and consumption, resulting in deviations from the desired frequency. The wind turbine power is related to the wind speed and expressed in watts given in equation 5.1.

$$P_w = \frac{1}{2}\rho A V^3 C_p(\lambda, \alpha)$$ (5.1)

where:

P_w = Power output (W)
ρ = Air density (kg/m^3)
A = Turbine blade's swept area (m^2)
V = Wind speed (m/s)
λ = Tip speed ratio
α = Pitch angle (degrees or radians)
C_p = Power coefficient (depends on λ and α)

Figure 5.1 shows the mathematical model of the proposed system, and Figure 5.2 shows the conventional power system model incorporating the wind source.

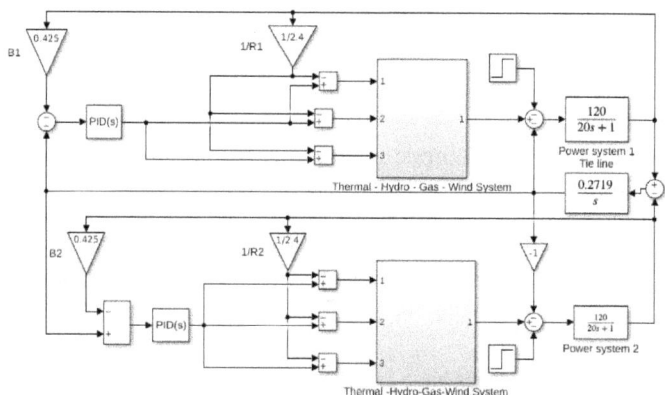

Figure 5.1 Interconnected power system with a renewable source (wind) transfer function model

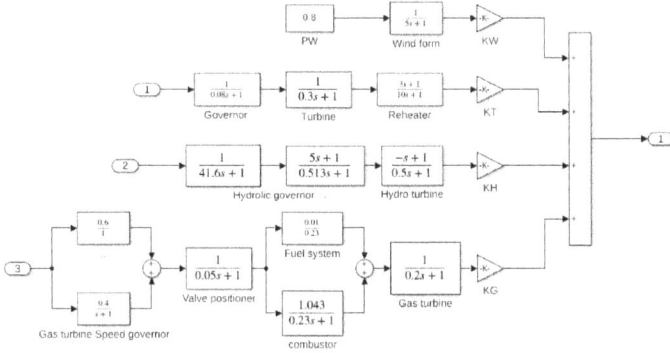

Figure 5.2 Wind energy source incorporating a multisource power system

Challenges of wind energy source implementation

When wind energy sources are incorporated into a multisource power system in multiple areas, several challenges, such as intermittency, variability, and grid stability, are present.

5.2 SIMULATION RESULTS AND DISCUSSION

The performance of the proposed ACO method-based PID controller against frequency deviation for a conventional power system incorporating wind energy is illustrated in this section. The PID controller design and objective function utilized in this book are explained in Chapter 2, Section 4.2. The different optimization methods (GA, PSO, FA, & BAT)-tuned PID controller gain parameters are presented in Table 5.1. A graphical value comparison of the fitness values is shown in Figure 5.3.

Table 5.1

Gain values of the GA-tuned, PSO-tuned, FA-tuned, BA-tuned, and ACO technique-tuned PID controller gain values with fitness values

Technique/ Gain Parameter	Kp1	Ki1	Kd1	Kp2	Ki2	Kd2	Fitness Value J
GA	0.5522	0.9726	0.4923	0.6947	0.9790	0.3277	5.7617
PSO	0.7158	0.9939	0.8271	0.2358	0.9998	0.0447	5.7461
FA	0.7000	0.9998	0.7482	0.7516	0.9998	0.2613	5.6426
BAT	0.6353	0.9993	0.9212	0.4373	0.9998	0.0123	5.6481
ACO	0.6511	0.9998	0.6479	0.5854	0.9998	0.2372	5.6142

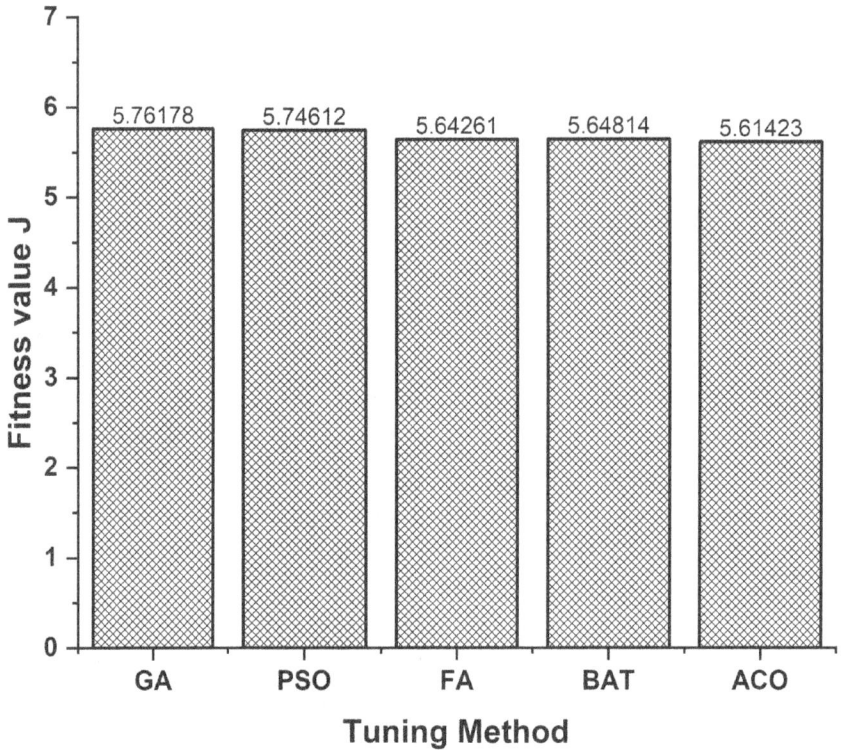

Figure 5.3   Bar chart evaluation of the fitness value J

5.2.1   PERFORMANCE ANALYSIS

The tuned parameters are utilized to demonstrate the dominance of the ACO method over frequency oscillation in the wind-incorporated investigated power system. The system response was analysed with a 1% load disturbance. The graphical response comparison for the area 1 and area 2 frequency differences is shown in Figures 5.4 and 5.5, the tie-line power difference is shown in Figure 5.6, and the ACE difference is shown in Figures 5.7 and 5.8. The numerical values from Figures 5.4 to 5.8 are reported in Tables 5.2 to 5.4.

By analyzing the system response, it is clear and effective that the ACO–PID controller provides a greater response than other optimization methods, such as the GA, PSO, FA, and BAT, in terms of settling time and peak shoots during emergency loading conditions compared with other technique-tuned controller parameters. The ACO technique-tuned controller takes 48 s and 50 s to settle frequency deviations in areas 1 and 2 and 24 s to settle deviations in tie-line power (del Ptie) between connected areas. Similarly, the area control error in areas 1 and 2 takes 45 s and 42 s, respectively, to settle the response under stable conditions. The time-domain

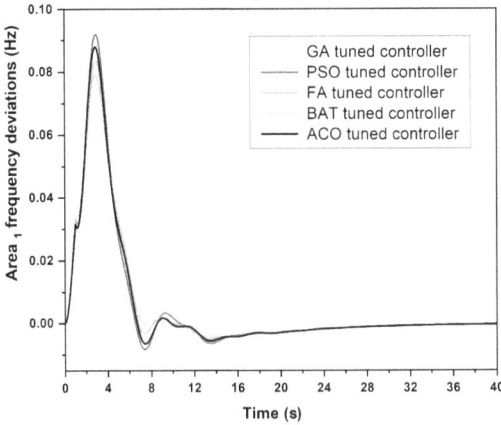

Figure 5.4   Assessment of frequency deviations in area 1

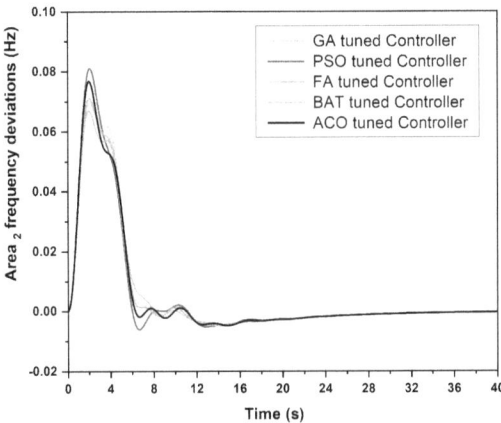

Figure 5.5   Assessment of frequency deviations in area 2

specification parameters are the settling time (Ts) and peak shoots (over (Os) and under (Us)) of each optimization technique, and comparisons are analysed in Chapter 5, Section 5.2.2.

5.2.2   TIME DOMAIN PARAMETERS OF SYSTEMS WITH DIFFERENT OPTI-MIZATION TECHNIQUE-TUNED PID CONTROLLERS

The mathematical values of the time domain specification parameters are listed in Tables 5.2, 5.3, and 5.4 for the system frequency in areas 1 and 2, del Ptie, and ACE in areas 1 and 2, respectively.

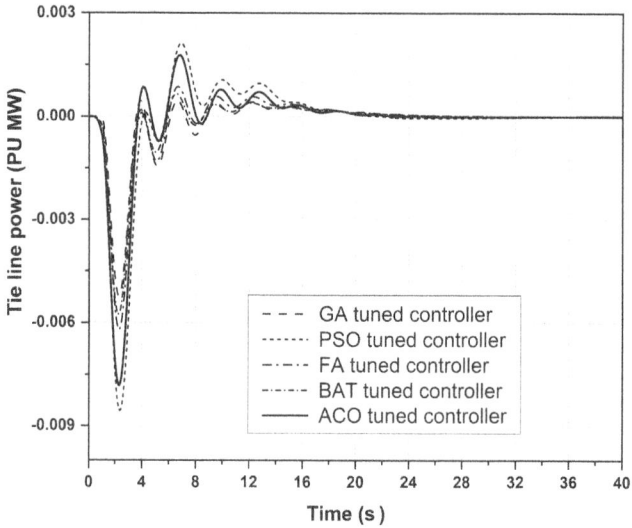

Figure 5.6   Assessment of tieline power flow deviations

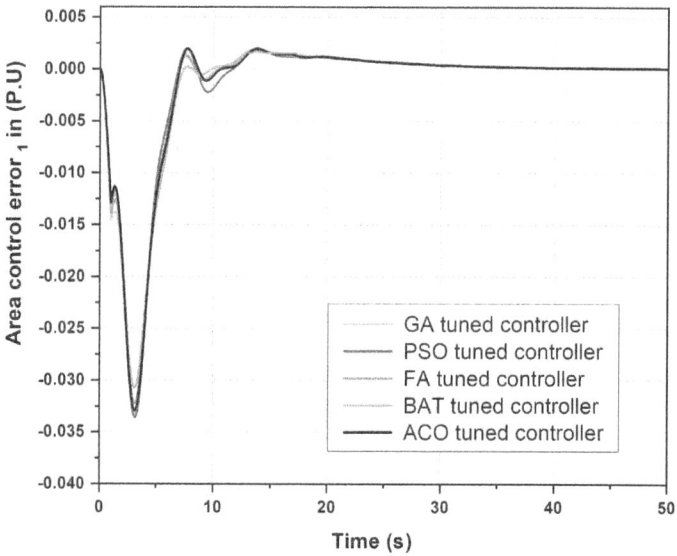

Figure 5.7   Assessment of the ACE in area 1

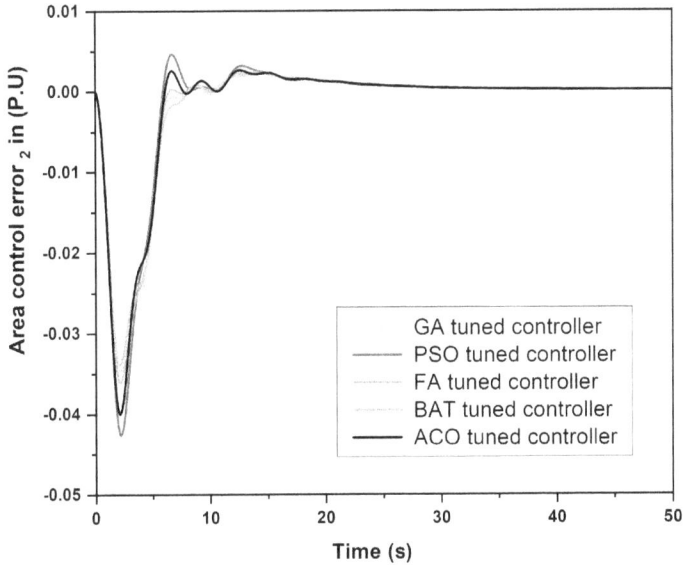

Figure 5.8   Assessment of the ACE in area 2

Table 5.2
Evaluation of the time-domain parameters with respect to the system frequency

Technique	ΔF_1			ΔF_2		
	TS (S)	OS (Hz)	US (Hz)	TS (S)	OS (Hz)	US (Hz)
GA	60	8.366×10^{-2}	5.071×10^{-3}	60	6.960×10^{-2}	5.049×10^{-3}
PSO	60	9.196×10^{-2}	8.227×10^{-3}	60	8.103×10^{-2}	6.122×10^{-3}
FA	58	7.907×10^{-2}	4.385×10^{-3}	55	6.689×10^{-2}	4.516×10^{-3}
BAT	50	8.789×10^{-2}	6.517×10^{-3}	53	7.671×10^{-2}	4.790×10^{-3}
ACO	48	8.366×10^{-2}	5.089×10^{-3}	50	7.108×10^{-2}	4.831×10^{-3}

Table 5.3
Evaluation of time-domain parameters in tie-line power flow

Technique	TS (S)	OS (pu MW)	US (pu MW)
GA	30	4.528×10^{-4}	5.227×10^{-3}
PSO	30	2.119×10^{-3}	8.560×10^{-3}
FA	30	6.592×10^{-4}	5.753×10^{-3}
BAT	25	1.779×10^{-3}	7.819×10^{-3}
ACO	24	8.656×10^{-4}	6.194×10^{-3}

Table 5.4

Evaluation of the time-domain parameters with respect to the ACE

Optimization Technique	ACE 1			ACE 2		
	TS (S)	OS (pu)	US (pu)	TS (S)	OS (pu)	US (pu)
GA	60	1.915×10^{-3}	3.312×10^{-2}	50	2.388×10^{-3}	3.462×10^{-2}
PSO	55	2.011×10^{-3}	3.356×10^{-2}	48	4.462×10^{-3}	4.257×10^{-2}
FA	50	1.607×10^{-3}	3.072×10^{-2}	45	2.188×10^{-3}	3.338×10^{-2}
BAT	48	1.975×10^{-3}	3.290×10^{-2}	44	2.625×10^{-3}	3.992×10^{-2}
ACO	45	1.718×10^{-3}	3.224×10^{-2}	42	2.387×10^{-3}	3.162×10^{-2}

The mathematical value analysis in Tables 5.2, 5.3, and 5.4 and the evaluation comparisons in the above section prove that the ACO method tuned controller settles the response in the shorter period (f1 = 48 s, f2 = 50 s, tie-line = 24 s, ACE1 = 45 s, ACE2 = 42 s). This demonstrates that the ACO PID controller takes less time to respond than the GA, PSO, FA, and BAT method-tuned controllers do. To show the supremacy of the ACO technique, a PID controller bar chart is plotted for different technique-tuned controller settling times, and it is depicted in Figures 5.9 to 5.13. Compared with that in Chapter 2, while incorporating wind energy sources, the controller takes more time to settle the frequency deviation. Wind energy is the nonlinear power source for the system.

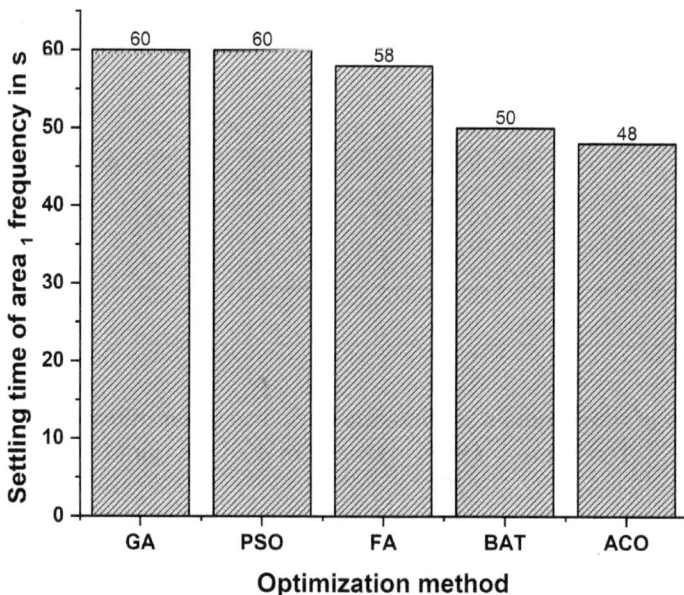

Figure 5.9 Bar chart evaluation of the settling time at the system frequency in area 1

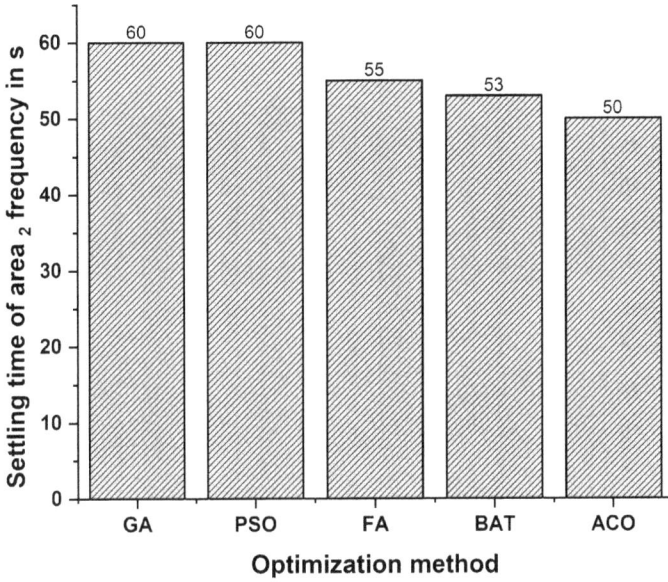

Figure 5.10   Bar chart evaluation of the settling time of the system frequency in area 2

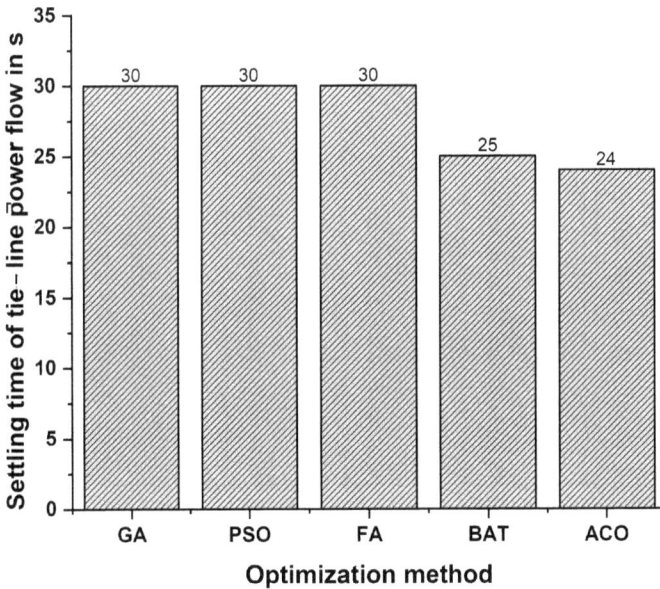

Figure 5.11   Bar chart evaluation of the tie-line power settling time

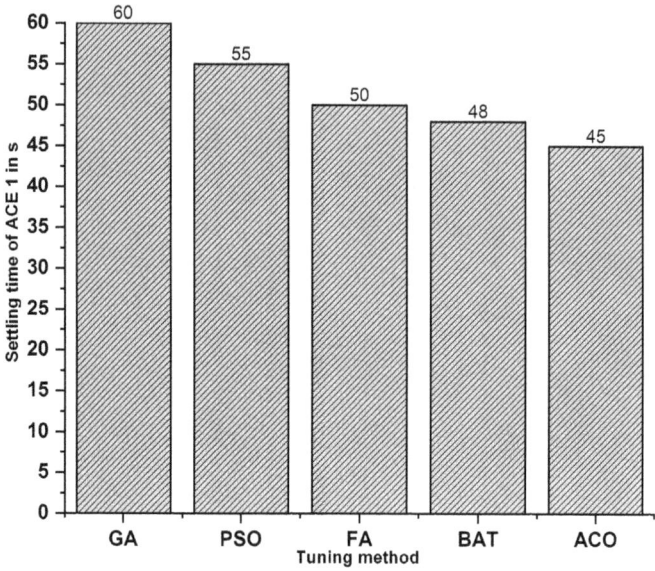

Figure 5.12   Evaluation of the settling time via a bar chart of the ACE in area 1

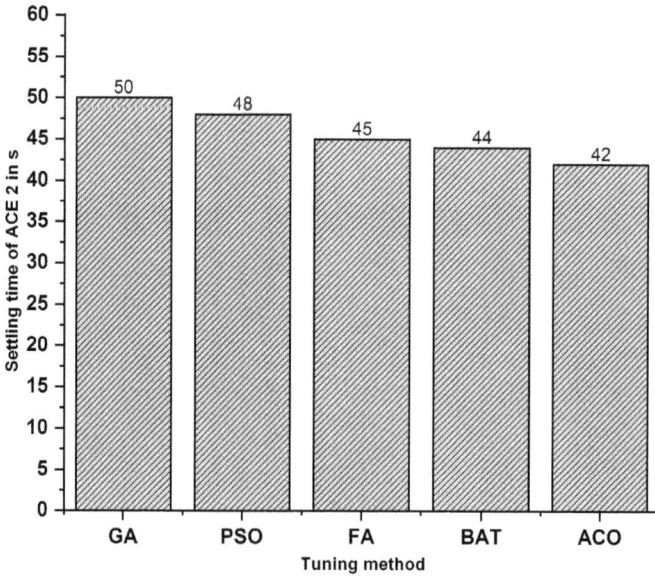

Figure 5.13   Evaluation of the settling time via a bar chart of the ACE in area 2

In Figures 5.9 to 5.13, a bar chart assessment of the settling time in areas 1 and 2, tie-line power flow, and ACE deviations demonstrated that the ACO–PID controller performed better than the other optimization methods, such as the GA, PSO, FA, and BAT. The improvement analysis of ACO over other optimization techniques is explained in Section 5.2.3.

5.2.3 ACO-TUNED PID CONTROLLER PERFORMANCE IMPROVEMENT OVER THE GA, PSO, FA, AND BA TECHNIQUE-TUNED PID CONTROLLERS IN SETTLING TIME

The performance enhancement of ACO–PID over other techniques, such as the genetic algorithm, particle swarm optimization technique, firefly algorithm, and bat algorithm, is evaluated in this section, with numerical values given in the table below and figures.

Table 5.5

Evaluation of the time-domain parameters with respect to the system frequency

Technique	ACO–PID controller improvement (%) in area 1 system frequency	ACO–PID controller improvement (%) in area 2 system frequency
ACO over GA	20	16
ACO over PSO	20	16
ACO over FA	16	8
ACO over BA	4	5

With respect to the performance improvement analysis in Table 5.5 and the bar chart evolution in Figures 5.14 and 5.15, effectively, the ACO-tuned controller delivers a better-controlled response in terms of minimal settling, which takes more time than the other controllers do.

Table 5.6

Evaluation of time-domain parameters in tie-line power flow

Technique	Improvement in Tie-Line Power Flow (%)
ACO over GA	24
ACO over PSO	24
ACO over FA	24
ACO over BA	4

On the basis of the performance enhancement analysis of ACO-tuned controller action, the numerical values are listed in Table 5.6, and the bar chart evolution is

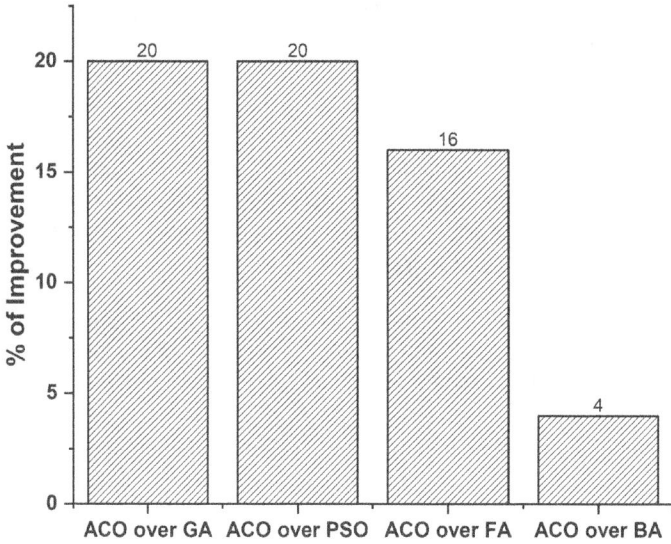

Figure 5.14   Evaluation of the settling time via a bar chart of the system frequency in area 1

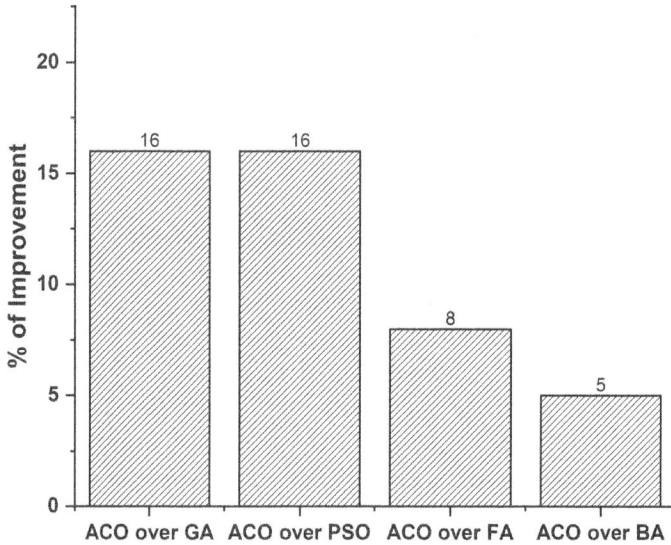

Figure 5.15   Evaluation of the settling time via a bar chart of the system frequency in area 2

shown in Figure 5.16, which clearly shows that the ACO-tuned controller offers a better-controlled response in terms of minimal settling, which takes more time than the other controllers do.

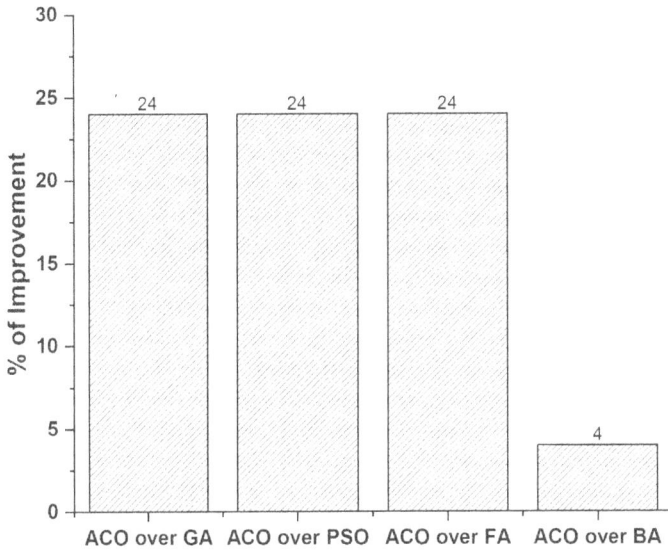

Figure 5.16   Evaluation of settling time via a bar chart for tie-line power flow

---

Table 5.7
Evaluation of the time-domain parameters in the area control error

Technique	ACO–PID controller improvement (%) in area 1 ACE	ACO–PID controller improvement (%) in area 2 ACE
ACO over GA	33	19
ACO over PSO	22	14
ACO over FA	11	7
ACO over BA	7	5

To prove the superiority of the ACO controller, numerical values of improvement are tabulated in Table 5.7, and the bar chart evolution in Figures 5.17 and 5.18 clearly shows that the ACO-tuned controller delivers a healthier controlled response. From the response analysis in Figures 5.4 to 5.8, the numerical values of the time-domain specification parameters and fitness values for each optimization technique and bar chart evaluation graphs of the settling time in Figures 5.9 to 5.13 prove that the ACO technique-tuned PID controller produced a superior response (minimal settling time and peak shoots) over the GA, PSO, FA, and BAT techniques optimized the controller performance in the emergency load demand situation in interlinked multiple areas with multiple sources incorporated with a wind source power system.

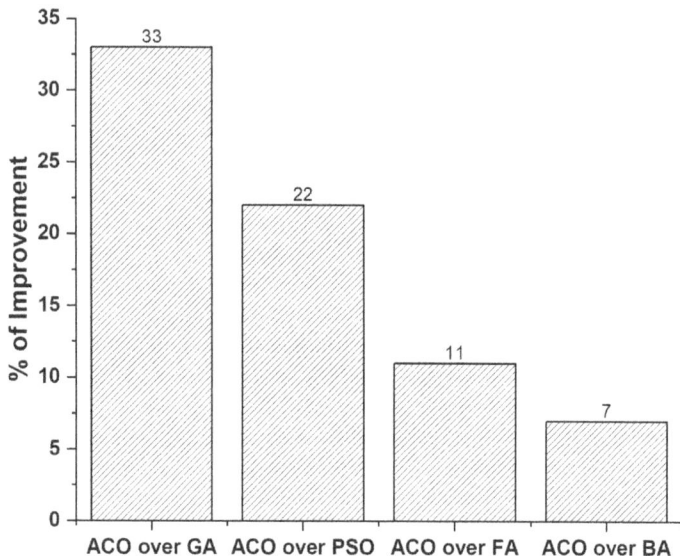

Figure 5.17   Evaluation of the settling time via a bar chart of the ACE in area 1

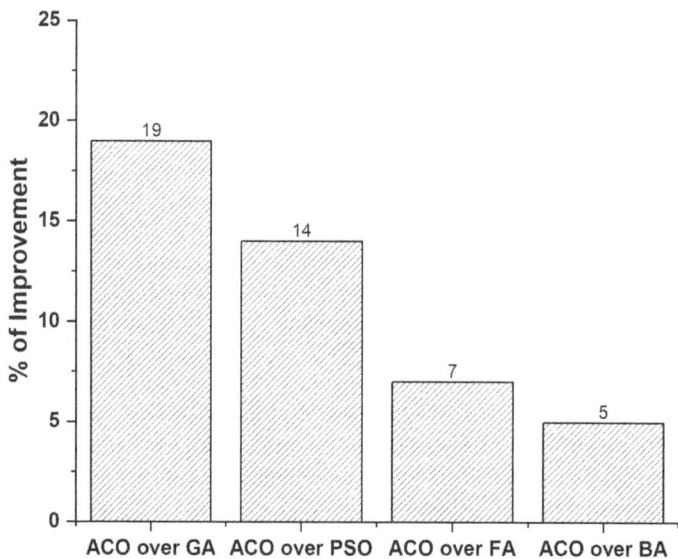

Figure 5.18   Evaluation of the settling time via a bar chart of the ACE in area 2

5.2.4   IMPACT ANALYSIS OF WIND ENERGY SOURCES IN MULTISOURCE MULTIAREA POWER SYSTEMS

When a wind energy source with a multiarea multisource power system is incorporated, stability issues such as late settling times and peak shoot changes arise. The exact impact of the wind source is analysed in this section. The graphical analysis of the areas 1 and 2 frequency difference, tie-line power, and ACE of areas 1 and 2 is shown in Figures 5.19 to 5.23. The numerical values from Figures 5.19 to 5.23 are given in Table 5.7.

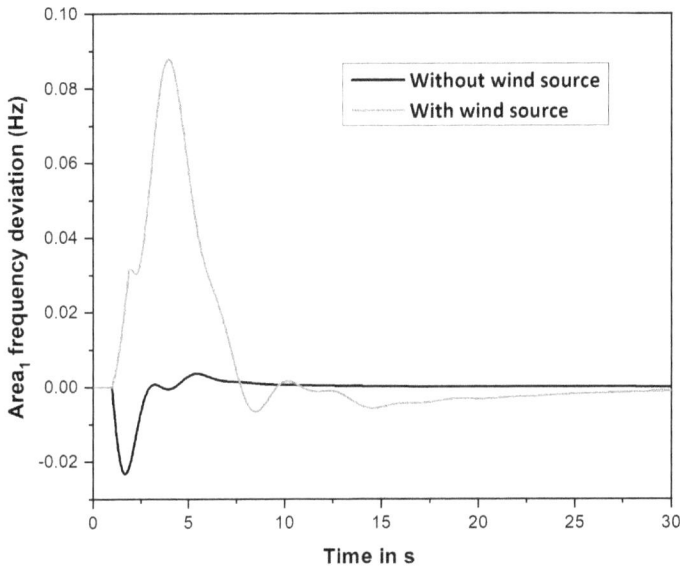

Figure 5.19 Impact analysis of the wind source frequency in area 1

The impact of the wind energy source in the multisource multiarea power system for LFC is analysed in this section. Figures 5.19 to 5.23 and Table 5.8 clearly present the impact of the wind source in the proposed system. When the wind energy source is incorporated with the proposed power system, the stability of the system is disturbed, and it takes time to settle the oscillation in system frequency and tieline power. The wind energy source is a nonlinear power source in the power system.

5.3 CONCLUSION

The proposed multisource and multiarea power system with a renewable energy source (wind)is investigated for LFC in this chapter. The ACO–PID controller has been suggested as a secondary controller to regulate system stability during sudden loading. Each in the investigated power system is identical, which consists of both conventional and renewable sources such as thermal, hydro, gas, and wind power sources. The proposed secondary controller gain parameters are tuned genetic algorithms, particle swarm optimization techniques, firefly algorithms, bat algorithms, and ant colony optimization techniques. During the optimization, the system is under 1% load demand. The proposed ACO–PID controller is a better secondary controller for the proposed system to maintain system stability at the time of

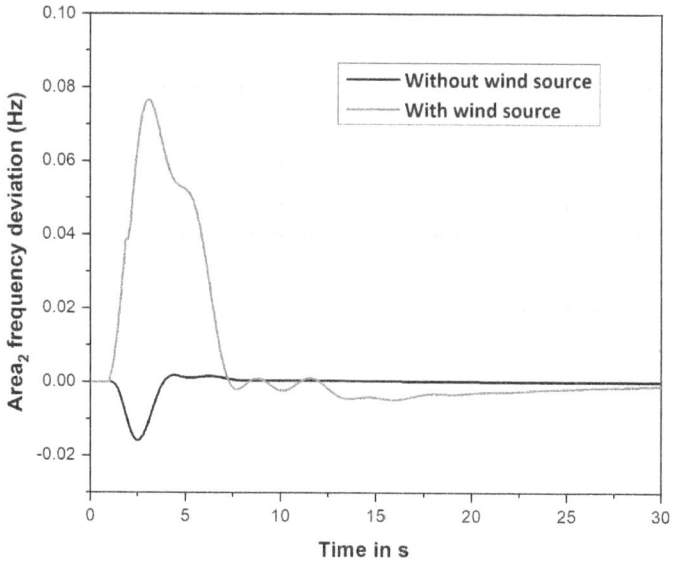

Figure 5.20   Impact analysis of the wind source frequency in area 2

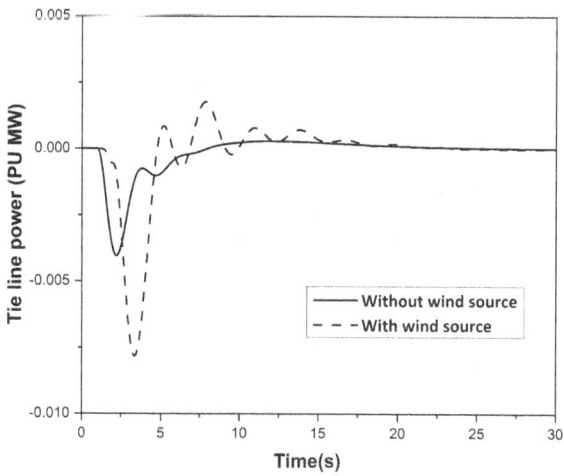

Figure 5.21   Impact analysis of the tie-line power of the wind source

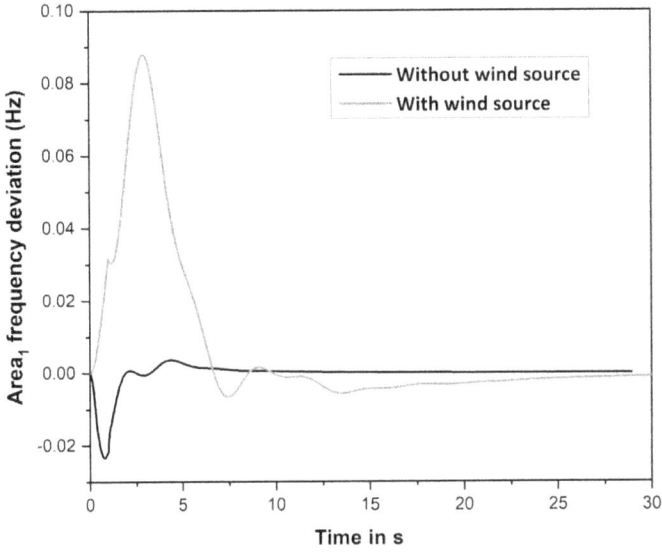

Figure 5.22   Impact analysis of the wind source in ACE 1

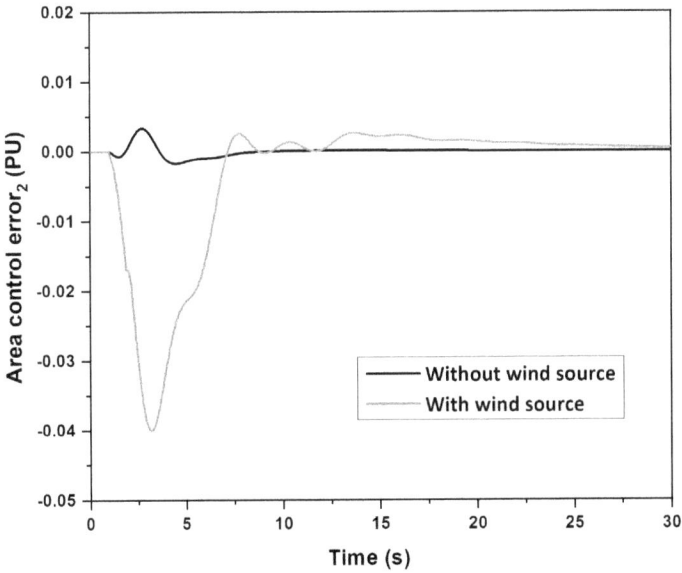

Figure 5.23   Impact analysis of the wind source in ACE 2

Table 5.8

Comparison of optimized parameters with and without wind integration

Optimized parameters/System		TS (S)	OS (Hz)	US (Hz)
Area 1 frequency	Without wind	17	3.6×10^{-3}	0.023
	With wind	48	8.366×10^{-2}	5.089×10^{-3}
Area 2 frequency	Without wind	27	1.8×10^{-3}	0.0155
	With wind	50	7.108×10^{-2}	4.831×10^{-3}
Optimized parameters/System		**TS (S)**	**OS (pu MW)**	**US (pu MW)**
Tie-line power	Without wind	29	3×10^{-4}	4×10^{-3}
	With wind	24	8.656×10^{-4}	6.194×10^{-3}
Optimized parameters/System		**TS (S)**	**OS (pu)**	**US (pu)**
ACE 1	Without wind	31	0.0127	1×10^{-3}
	With wind	45	1.718×10^{-3}	3.224×10^{-2}
ACE 2	Without wind	25	3.4×10^{-3}	1.7×10^{-3}
	With wind	42	2.387×10^{-3}	3.162×10^{-2}

uncertainty loading. The time domain specification of the chapter provides a clear dominance of ACO over other optimization methods in terms of fast settling of system frequency oscillation and minimal peak values. ACO gives a quick settling response (F1= 48 s, f2=50 s, tie-line power =24 s, ACE1=45 s, and ACE2=42 s).

6 Load frequency control of the multiarea multisource power system with renewable energy and energy storage unit

This chapter addresses the system frequency stability analysis for a multiarea multi-source power system with renewable and energy storage units. In the previous chapter, the system without energy storage units was analysed. In this chapter, two energy storage units, namely, hydroaquic electrolyzers (HAEs) and fuel cells (FCs), are integrated with the previously developed mathematical model. This chapter is structured as follows: the energy storage unit incorporating the system is described in Section 6.1. Section 6.2 presents a comparison of the simulation results, and Section 6.3 presents the conclusions of the chapter.

6.1 ENERGY STORAGE UNITS INCORPORATED INTO THE INVESTIGATED SYSTEM

In this chapter, two different energy storage units, the HAE and FC, are incorporated. The major role of the energy storage units in this power system is providing support to the power system during unilinear power production from renewable energy sources. The key points concerning HAE and FC are as follows.

6.1.1 HYDROGEN ELECTROLYSER

Energy Storage using Hydrogen: Electrolysis produces hydrogen, which may be stored and utilized as a source of energy. It may be kept in tanks and utilized with combustion engines or fuel cells to generate power at a later time. The benefits of the HAE are flexibility and scalability.
   Categories of Electrolyzers:

- The Membrane for Proton Exchange (PEM) Electrolyzers have great efficiency and quick reaction times.

DOI: 10.1201/9781003661153-6

- Alkaline Electrolyzers – Less expensive and employed in large-size industries with high technology.

- Solid oxide electrolysers (SOEs) can work at high temperatures with the highest efficiency.

Energy Storage using Hydrogen: Electrolysis produces hydrogen, which may be stored and utilized as a source of energy. It may be kept in tanks and utilized with combustion engines or fuel cells to generate power at a later time. The benefits of the HAE are flexibility and scalability.

- Flexibility: Hydrogen has a long storage life and may be used as needed, which helps to maintain a balance between supply and demand.
- Scalability: Systems for storing hydrogen can be expanded for more extensive uses.
- Renewable Integration: This makes it possible to store extra renewable energy for later use, such as solar or wind power.

Application:

- Grid balancing: Excess power from renewable sources is stored and discharged during times of high demand.
- Transport: One way to decarbonize the transportation industry is by using stored hydrogen as fuel for hydrogen fuel cell automobiles.
- Applications in industry: Hydrogen is applied as a feedstock or energy source in a range of industrial operations.

Challenges:

- Cost: Hydrogen production, storage, and conversion are quite expensive.
- Infrastructure: Infrastructure for storage and delivery is not in demand.
- Efficiency: During the conversion of electricity into hydrogen and back to electricity, the energy lost in the process is high.

Integration with renewable energy: Systems for storing and electrolyzing hydrogen are especially good at combining renewable energy sources such as wind and solar power. These systems decrease the unpredictability and intermittency related to renewable energy sources by turning excess renewable electricity into hydrogen. Energy security and resilience can be improved by converting stored hydrogen back into power at times when the renewable energy supply is scarce or demand is high. The mathematical expression of the HAE is given in equation 6.1

$$G_{HAE} = \frac{K_{HAE}}{1 + s.T_{HAE}} \tag{6.1}$$

where

- K_{HAE} – gain constant of the HAE.
- T_{HAE} – Time constant of the HAE

Future scope of HAE: Hydrogen electrolysers and energy storage devices are expected to be vital components of the worldwide shift toward sustainable energy sources as technology advances. There are ongoing initiatives to lower expenses, increase productivity, and develop infrastructure, which are backed by global private sector investments and government regulations.

6.1.2 FUEL CELLS

Fuel cells are becoming widely incorporated into power systems as a dependable and effective way to produce energy. It has low emissions, great efficiency, and adaptability for various applications, from large-scale industrial power plants to domestic power supplies. An outline of fuel cell integration into power systems is provided below:

Fuel cells in a power system:
The role of the fuel cell in the power system is described below.

Dispersed Production: Distributed generation systems, in which power is produced at the point of use instead of at a sizable, centralized power plant, can make use of fuel cells. Energy dependability is improved, and transmission losses are decreased as a result. They offer a reliable power source to homes, companies, and isolated areas, making them perfect for peak shaving, off-grid applications, and backup power.

Combined heat and power (CHP): A technique known as cogeneration, or combined heat and power, allows fuel cells to simultaneously create usable heat and electricity (CHP). In doing so, the heat produced during the creation of electricity is used, increasing total energy efficiency. The benefits of CHP systems may reach efficiencies of over 80%, and they are appropriate for settings such as hospitals, universities, and industrial sites that need both heating and power.

Integration of renewable energy: Hydrogen production is the reverse process of fuel cells, which can function as electrolysers to generate hydrogen from surplus renewable energy. When renewable energy sources are insufficient, this hydrogen may be stored and then utilized in fuel cells to produce power. In addition, for grid stability, fuel cells, especially when combined with intermittent renewable energy sources such as solar and wind, help maintain grid stability by balancing supply and demand.

Basic Components:

- Anode: The negative side where hydrogen gas (H_2) is supplied.
- Cathode: The positive side where oxygen (O_2) is supplied, usually from the air.
- Electrolyte: A medium that allows ions to move between the anode and cathode.

Operation in a fuel cell: Gaseous hydrogen is supplied to the anode, where a catalyst helps split the gas into protons (H^+) and electrons (e^-). While the protons travel through the electrolyte to the cathode, the electrons flow through an external circuit to produce electricity. Heat and water are produced at the cathode when oxygen reacts with protons and electrons.

Fuel Cell Benefits for Power Systems

- **Efficiency:** Compared with traditional combustion-based power production, fuel cells are more efficient since they immediately transform chemical energy into electrical energy.
- **Low Emissions:** These products release very little, mostly heat and water, which helps to clean the air and reduce greenhouse gas emissions.
- **Quiet operation:** Because fuel cells run silently, they are a good choice for metropolitan settings where noise pollution is an issue.
- **Modularity and Scalability:** Fuel cell systems are adaptable to a wide range of applications because of their flexibility in scaling up or down to suit specific power demands.

Obstacles and Things to Think About

- **Cost:** This is a more costly power production method than conventional power production methods. This is because the electrodes are much more costly.
- **The Infrastructure for Hydrogen Production, Storage, and Distribution:** One of the primary obstacles to the broad use of hydrogen fuel is the availability of hydrogen fuel and the necessary infrastructure.
- **Durability and Lifespan:** To prolong their useful life and lower maintenance expenses, fuel cells need to be designed and manufactured with new materials and techniques.

Power System Applications

- **Residential Power Supply:** Fuel cell systems installed in homes may provide electricity and heat, providing a cost-effective and environmentally friendly substitute for conventional power sources.
- **Commercial & Industrial Power:** Fuel cells may provide dependable power to businesses and industrial facilities, minimizing dependency on the electrical grid and increasing energy security.
- **Large-Scale Power Plants:** By producing electricity for the grid, fuel cell power plants may help achieve clean energy targets and reduce emissions from conventional power plants.

The mathematical expression of the fuel cell is given in Equation 6.2.

$$G_{FC} = \frac{K_{FC}}{1 + T_{FC}s} \tag{6.2}$$

where

K_{FC} - gain constant of the fuel cell and T_{FC} - time constant of the fuel cell.

Prospects for the Future Fuel cells are predicted to play an increasingly important role in power systems as technology develops and their cost continues to decline. Fuel cells play a significant role in the global energy landscape as concerns about climate change and the demand for sustainable energy alternatives increase. Encouraging regulations and investments in hydrogen infrastructure will be essential in accelerating the use of fuel cell technology in power systems.

6.2 SIMULATION RESULTS AND DISCUSSION

The proposed ACO method-based PID controller response over frequency oscillation for energy storage units (HAE and FC) and a conventional power system incorporating wind energy is illustrated in this section. The PID controller design and objective function utilized in this book are explained in Chapter 4, Section 4.3. The different optimization methods (GA, PSO, FA, & BAT)-tuned PID controller gain parameters are presented in Table 6.1. A graphical comparison of the fitness values is shown in Figure 6.1.

Table 6.1

Gain values of the GA-tuned, PSO-tuned, FA-tuned, BA-tuned and ACO technique-tuned PID controller gain values with fitness values

Technique/ Gain Parameter	Kp1	Ki1	Kd1	Kp2	Ki2	Kd2	Fitness Value J
GA	0.6358	0.9451	0.4587	0.6619	0.9880	0.3502	0.8958
PSO	0.6584	0.9976	0.2338	0.9998	0.9977	0.8015	0.6751
FA	0.9224	0.9998	0.1686	0.9809	0.9999	0.3301	0.5306
BAT	0.9460	0.9988	0.1047	0.9957	0.9975	0.0513	0.4789
ACO	0.9993	0.9998	0.0113	0.9996	0.9999	0.0343	0.4629

6.2.1 RESPONSE ANALYSIS

The optimized (GA, PSO, BAT, FA, and ACO) controller gain parameters are implemented in the investigated power to demonstrate the supremacy of the ACO method against frequency oscillation. The performance is analysed by applying 1% load demand. A graphical response comparison for areas 1 and 2 is shown in Figures 6.2 and 6.3, the tie-line power difference is shown in Figure 6.4, the ACE difference is shown in Figures 6.5 and 6.6, and the numerical values from Figures 6.2 to 6.6 are reported in Tables 6.2 to 6.3.

The detailed response analysis is shown in Figures 6.2 to 6.8. The dominance of the proposed ACO–PID controller is greater than that of the other optimization methods utilized in the work, such as the GA, PSO, FA, and BAT. The ACO–PID

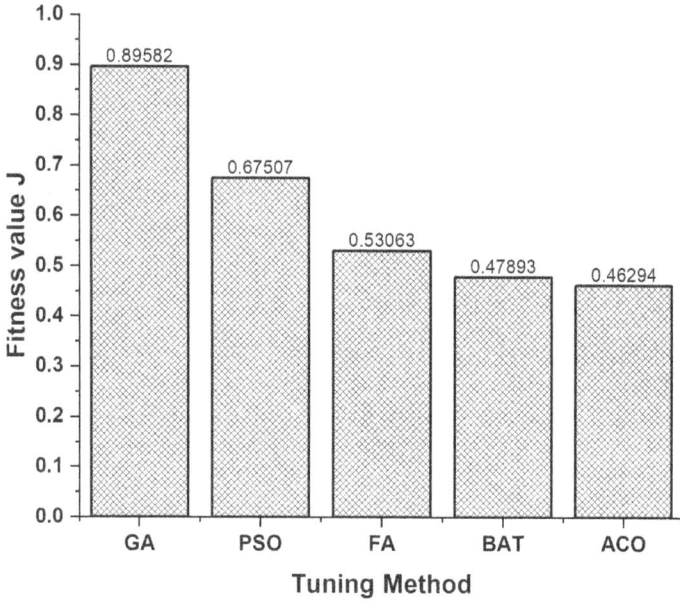

Figure 6.1 Bar chart of fitness value J

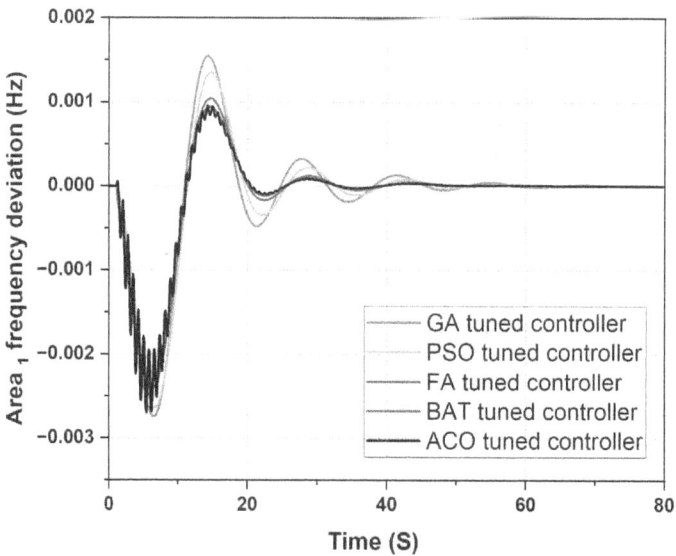

Figure 6.2 Assessment of frequency deviations in area 1

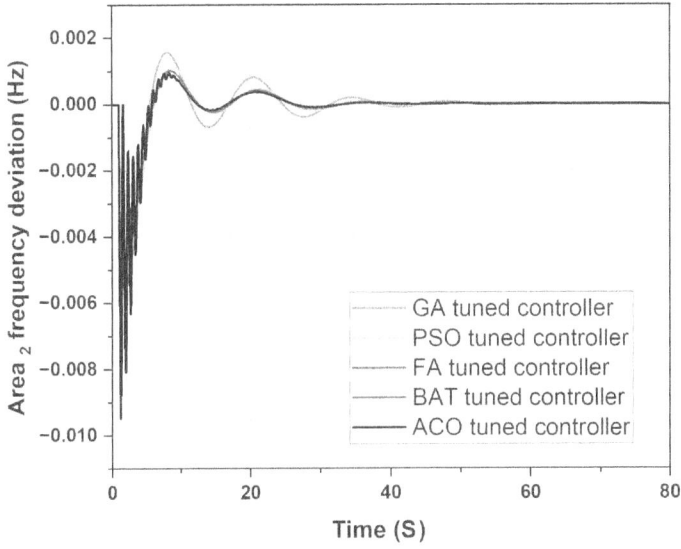

Figure 6.3   Assessment of frequency deviations in area 2

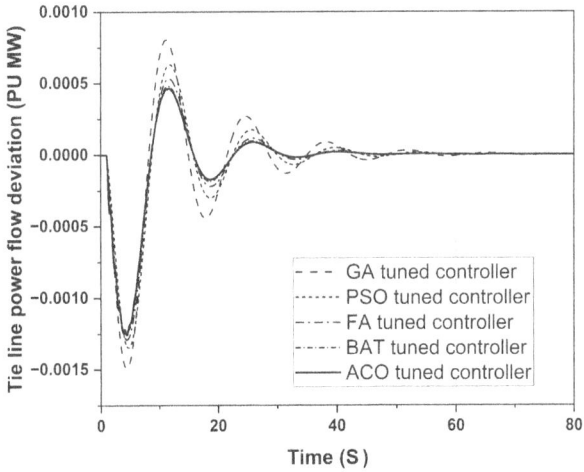

Figure 6.4   Assessment of Tie-line power flow deviations

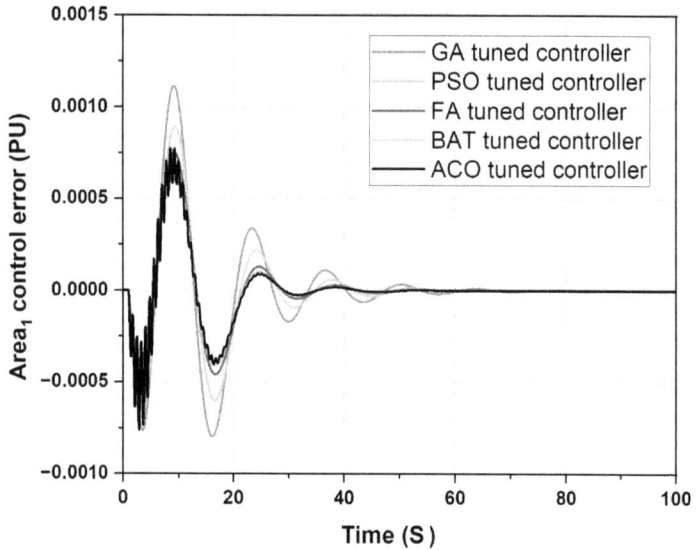

Figure 6.5   Assessment of the ACE in area 1

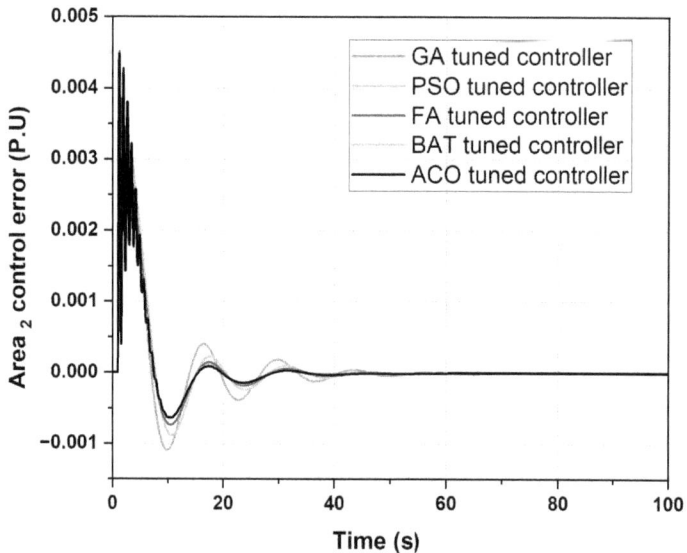

Figure 6.6   Assessment of the ACE in area 2

Table 6.2

Evaluation of the time domain parameters with respect to the system frequency

Optimization Technique	del F1			del F2		
	TS (S)	OS (Hz)	US (Hz)	TS (S)	OS (Hz)	US (Hz)
GA	62	1.548×10^{-3}	2.743×10^{-3}	59	1.573×10^{-3}	5.573×10^{-3}
PSO	60	1.357×10^{-3}	2.663×10^{-3}	55	1.119×10^{-3}	4.554×10^{-3}
FA	50	1.044×10^{-3}	2.412×10^{-3}	45	1.020×10^{-3}	5.598×10^{-3}
BAT	48	9.448×10^{-4}	2.360×10^{-3}	40	9.191×10^{-4}	9.103×10^{-3}
ACO	47	9.598×10^{-4}	2.703×10^{-3}	36	9.636×10^{-4}	9.473×10^{-3}

Table 6.3

Evaluation of time domain parameters in tie-line power flow

Optimization Technique	TS (S)	OS (pu MW)	US (pu MW)
GA	62	8.014×10^{-4}	1.497×10^{-3}
PSO	60	6.323×10^{-4}	1.347×10^{-3}
FA	46	5.317×10^{-4}	1.290×10^{-3}
BAT	45	4.709×10^{-4}	1.263×10^{-3}
ACO	44	4.643×10^{-4}	1.257×10^{-3}

controller settles the oscillation in the system frequency in areas 1 and 2, and the tie-line power flows are 47 s, 36 s, and 44 s, respectively, during the unexpected loading situations. Additionally, the ACO–PID controller response in the area control error is faster than that of the other techniques, and the ACE 1 and ACE 2 oscillations are set at 46 s and 43 s. Apart from the fast-settling time, all other time domain parameters, such as peak shoots (over (Os) and under (Us)), are also less than those of the other optimization methods. A detailed numerical comparison of the above response is presented in Section 6.2.2.

6.2.2   TIME-DOMAIN PARAMETERS OF SYSTEMS WITH DIFFERENT OPTIMIZATION TECHNIQUE-TUNED PID CONTROLLERS

The mathematical values of the time domain specification parameters are listed in Tables 6.2, 6.3, and 6.4 for the system frequency in areas 1 & 2, del Ptie, and ACE in areas 1 and 2, respectively.

The mathematical analysis in the above Tables 6.2, 6.3, and 6.4 and the performance comparisons in the above section prove that the ACO–PID controller settles the response in less time (f1 = 47 s, f2 = 36 s, tie-line = 44 s, ACE1 = 46 s, ACE2 = 43 s). The ACO–PID controller settles the oscillation in less time than the GA, PSO, FA, and BAT method-tuned controllers do. To show the supremacy of the ACO technique, a PID controller bar chart is plotted for different technique-tuned controller settling times, and it is depicted in Figures 6.7 to 6.11. When the energy storage units are in some time, the system frequency oscillation settles faster than previously because the system complexity in some cases requires more time.

Table 6.4

Evaluation of the time domain parameters in the area control error

Optimization Technique	ACE 1			ACE 2		
	TS (S)	OS (pu)	US (pu)	TS (S)	OS (pu)	US (pu)
GA	80	1.113×10^{-3}	7.952×10^{-4}	65	3.265×10^{-3}	1.090×10^{-3}
PSO	58	8.967×10^{-4}	6.286×10^{-4}	55	2.889×10^{-3}	8.850×10^{-4}
FA	51	7.551×10^{-4}	5.619×10^{-4}	46	2.839×10^{-3}	7.394×10^{-4}
BAT	48	7.129×10^{-4}	5.028×10^{-4}	44	4.314×10^{-3}	6.516×10^{-4}
ACO	46	7.751×10^{-4}	7.600×10^{-4}	43	4.510×10^{-3}	6.342×10^{-4}

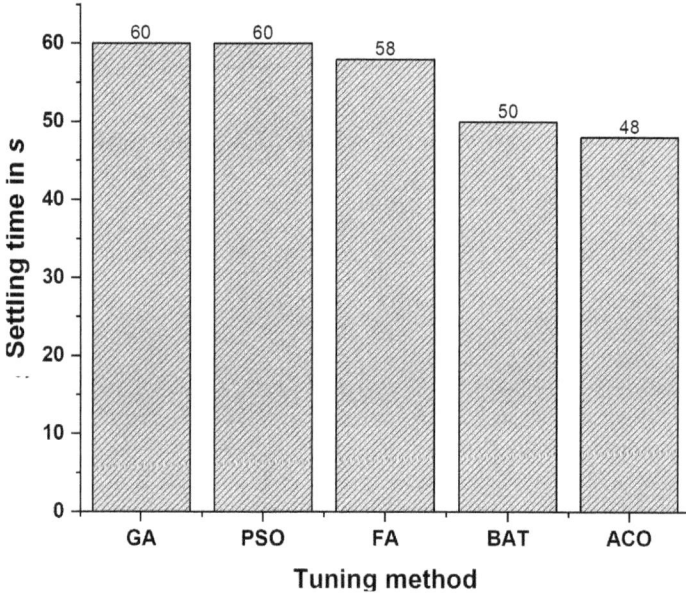

Figure 6.7   Bar chart evaluation of the settling time at the system frequency in area 1

In Figures 6.7 to 6.11, a bar chart assessment of the settling time in areas 1 and 2, tie-line power flow, and ACE deviations demonstrated that the ACO–PID controller performed better than the other optimization methods, such as the GA, PSO, FA, and BAT. The improvement analysis of ACO over other optimization techniques is explained in Section 6.2.3.

6.2.3   ACO-TUNED PID CONTROLLER PERFORMANCE IMPROVEMENT OVER THE GA, PSO, FA, AND BA TECHNIQUE-TUNED PID CONTROLLERS IN SETTLING TIME

The ACO–PID superiority in terms of percentage over other techniques, such as the genetic algorithm, particle swarm optimization technique, firefly algorithm, and bat algorithm, is evaluated in this section, with numerical values given in the table below and figures.

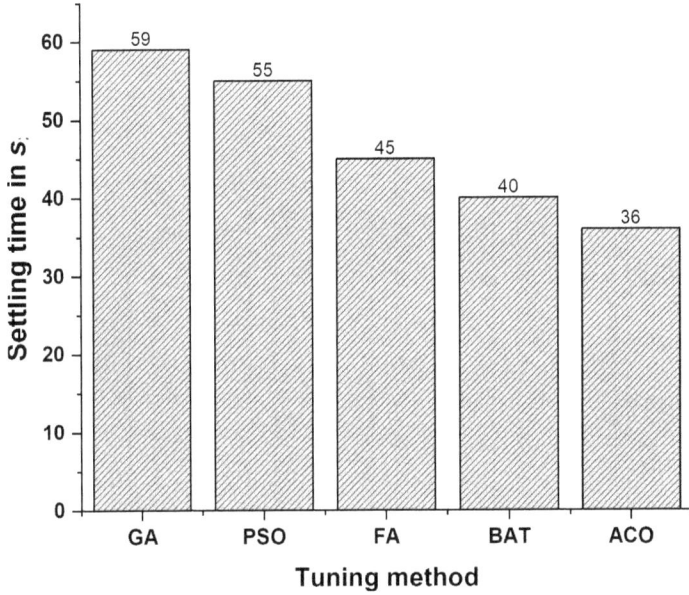

Figure 6.8   Bar chart evaluation of the settling time at the system frequency in area 2

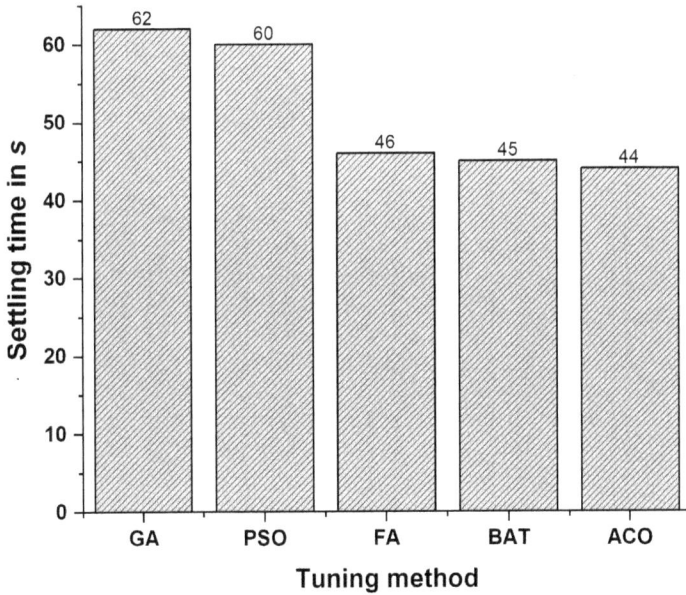

Figure 6.9   Bar chart evaluation of the tie-line power settling time

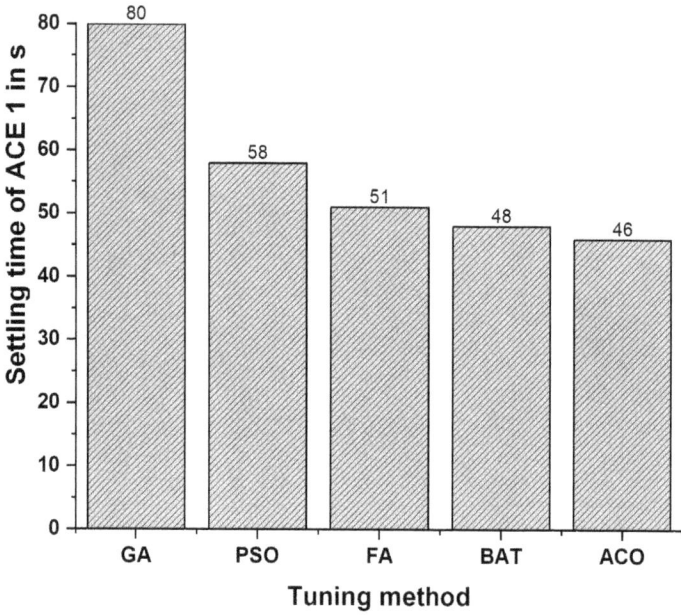

Figure 6.10   Evaluation of the settling time via a bar chart of the ACE 1

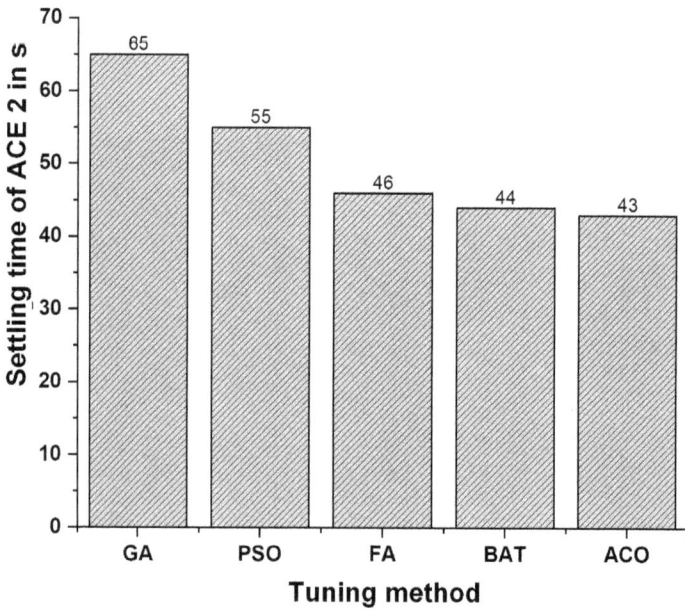

Figure 6.11   Evaluation of the settling time via a bar chart of the ACE 2

Table 6.5

Evaluation of the time-domain parameters with respect to the system frequency

Technique	Improvement in Area 1 Frequency (%)	Improvement in Area 2 Frequency (%)
ACO over GA	24	38
ACO over PSO	21	34
ACO over FA	6	20
ACO over BA	2	11

Table 6.6

ACO–PID Controller Improvement (%) in Tie-Line Power Flow

Technique	Improvement in Tie-Line Power Flow (%)
ACO over GA	29
ACO over PSO	27
ACO over FA	5
ACO over BA	2

Table 6.7

ACO–PID Controller Improvement (%) in ACE of Area 1 and Area 2

Technique	Improvement in Area 1 ACE (%)	Improvement in Area 2 ACE (%)
ACO over GA	43	34
ACO over PSO	21	22
ACO over FA	10	7
ACO over BA	4	2

The performance improvement analysis from Table 6.5 and the bar chart evolution in Figures 6.12 and 6.13 effectively prove that the ACO–PID controller gives a better-controlled response in terms of minimal settling, which takes more time than the other controllers do.

On the basis of the performance enhancement analysis of ACO-tuned controller action, the numerical values are listed in Table 6.6, and the bar chart evolution is shown in Figure 6.14, which clearly shows that the ACO-tuned controller offers a better-controlled response in terms of minimal settling, which takes more time than the other controllers do.

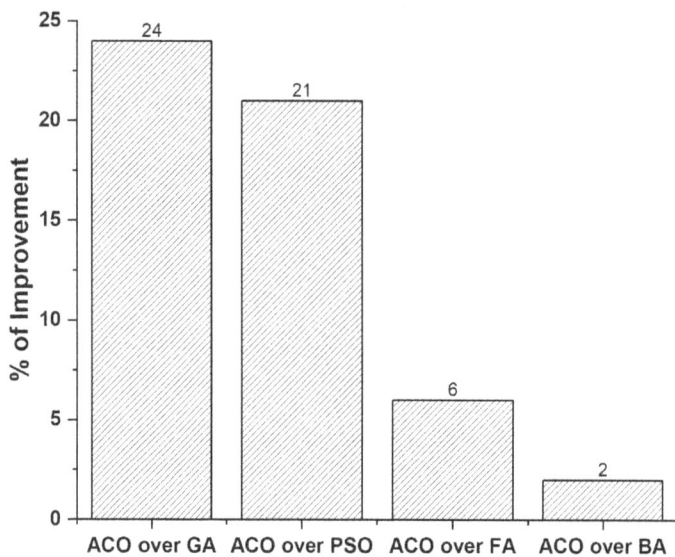

Figure 6.12   Evaluation of the settling time via a bar chart of the system frequency in area 1

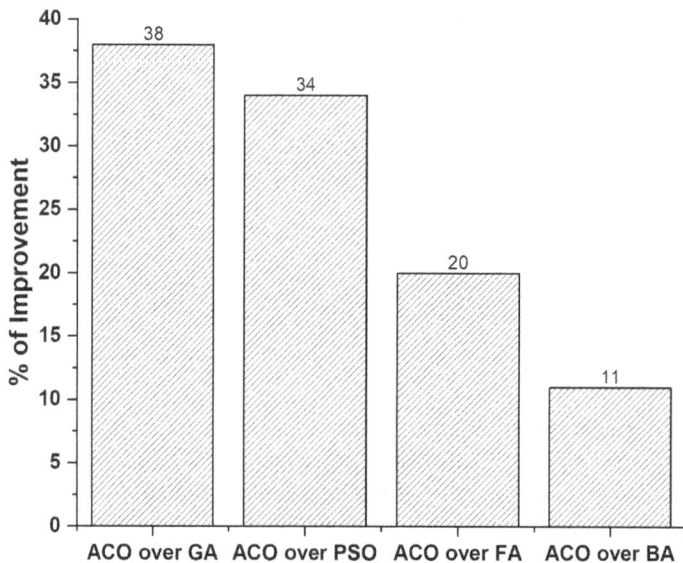

Figure 6.13   Evaluation of the settling time via a bar chart of the system frequency in area 2

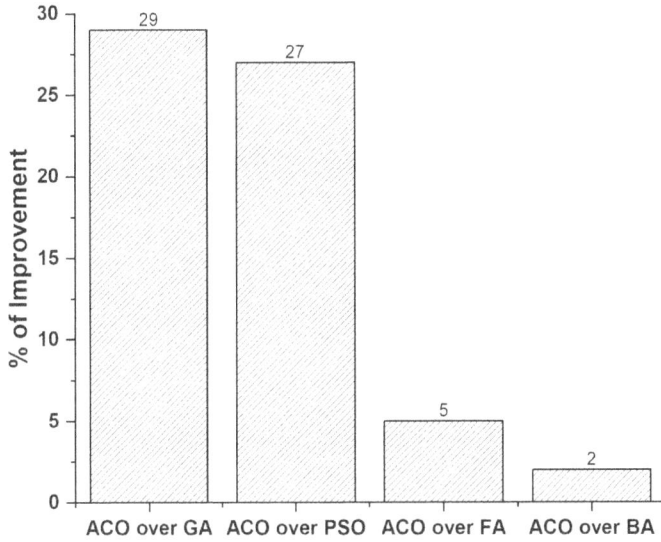

Figure 6.14   Evaluation of settling time via a bar chart for tie-line power flow

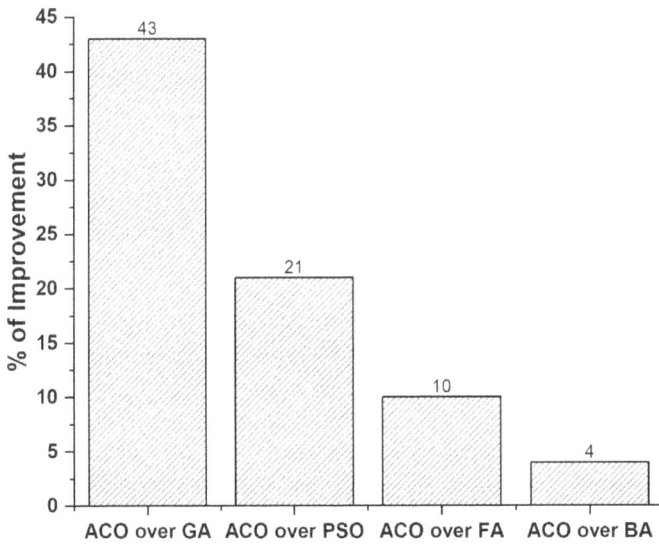

Figure 6.15   Evaluation of the settling time via a bar chart of the area control error in area 1

The superiority of the ACO–PID controller in terms of the numerical values of improvement is tabulated in Table 6.7, and the bar chart evolution in Figures 6.15 and 6.16 clearly shows that the ACO-tuned controller delivers a healthier controlled response.

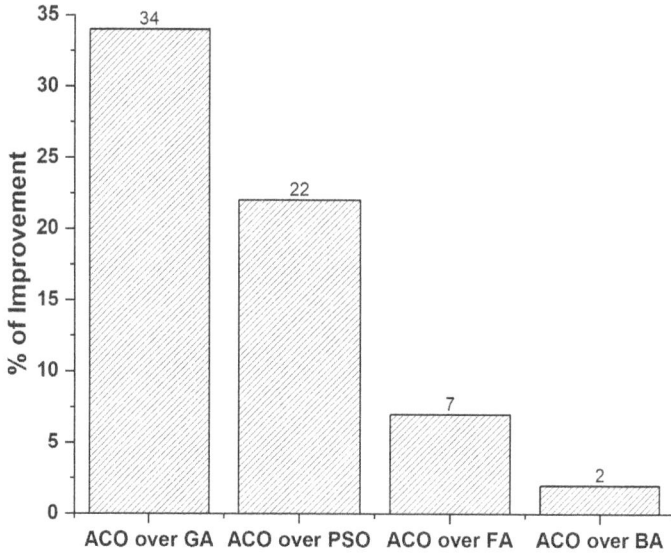

Figure 6.16 Evaluation of the settling time via a bar chart of the area control error in area 2

6.2.4   IMPACT ANALYSIS OF ENERGY STORAGE UNITS IN MULTISOURCE MULTIAREA POWER SYSTEMS

While energy storage units with renewable energy-based multiarea multisource power systems are included, stability issues such as settling time and peak shoot changes are controlled by the backup power from the energy storage units. The exact impact of energy storage is analysed in this section. The graphical analysis of the areas 1 and 2 frequency difference, tie-line power, and ACE of areas 1 and 2 is shown in Figures 6.17 and 6.18. The numerical values from Figures 6.17 and 6.18.

The impact of the energy storage units in the multisource multiarea with a renewable energy source power system for LFC is analysed in this section. Figures 6.17 to 6.21 and Table 6.8 clearly demonstrate the impact of the energy storage units in the proposed system. While incorporating the ESUs with the proposed power system, the system frequency oscillation is controlled, and uncertainty from the renewable sources is taken care of by the ESUs. The oscillations in the frequency in areas 1 and 2, the tie-line, ACE 1, and ACE 2 are settled quickly because of the energy storage units.

6.3   CONCLUSION

The proposed renewable energy source (wind)-based multisource and multiarea energy storage unit power system is investigated for LFC in this chapter. The ACO–PID controller has been suggested as a secondary controller to control and monitor system

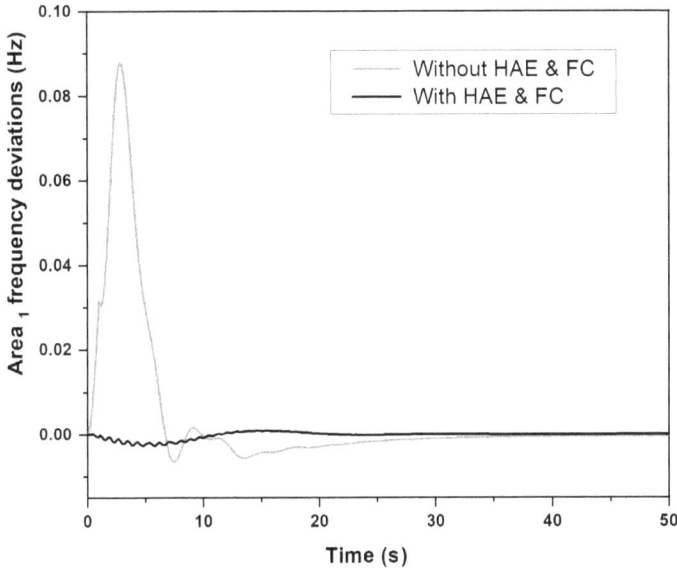

Figure 6.17 Impact analysis of energy storage units in area 1 frequency

Table 6.8
Comparison of optimized parameters with and without wind integration

System	Condition	TS (S)	OS (Hz)	US (Hz)
Area 1 frequency	With wind	47	9.598×10^{-4}	2.703×10^{-3}
	Without wind	48	8.366×10^{-2}	5.089×10^{-3}
Area 2 frequency	With wind	36	9.636×10^{-4}	9.473×10^{-3}
	Without wind	50	7.108×10^{-2}	4.831×10^{-3}
Tie-line power	With wind	44	4.643×10^{-4}	1.257×10^{-3}
	Without wind	24	8.656×10^{-4}	6.194×10^{-3}
ACE 1	With wind	46	7.751×10^{-4}	7.600×10^{-4}
	Without wind	45	1.718×10^{-3}	3.224×10^{-2}
ACE 2	With wind	43	4.510×10^{-3}	6.342×10^{-4}
	Without wind	42	2.387×10^{-3}	3.162×10^{-2}

stability during sudden loading. Each in the investigated power system is identical, which consists of both conventional and renewable sources such as thermal, hydro, gas, and wind power sources with energy storage units such as HAE and FC. The proposed secondary controller gain parameters are tuned genetic algorithms, particle swarm optimization techniques, firefly algorithms, bat algorithms, and ant colony

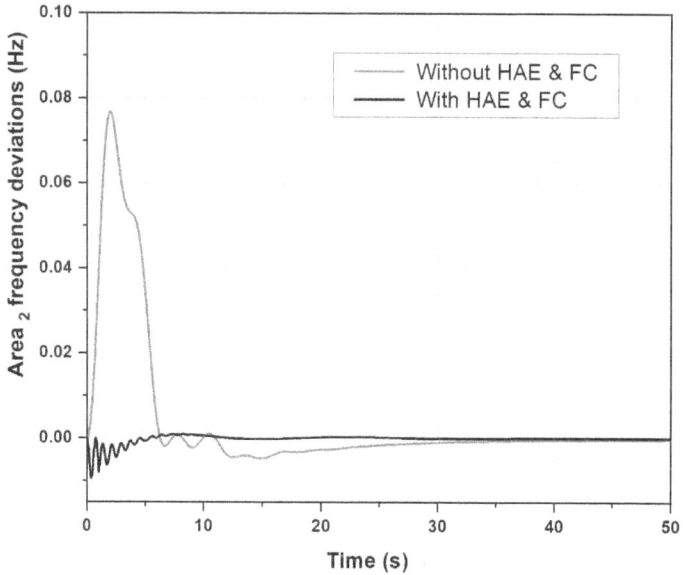

Figure 6.18   Impact analysis of energy storage units in area 2 frequency

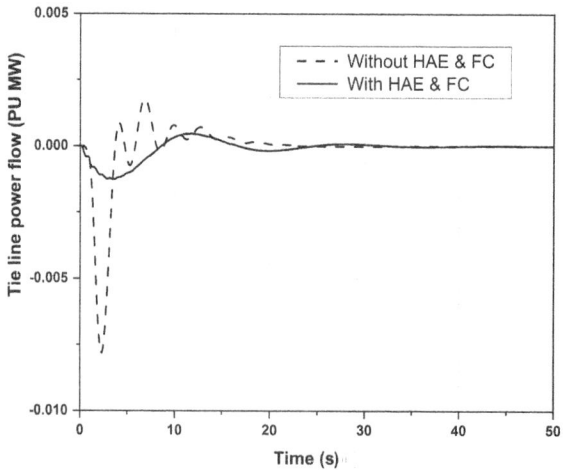

Figure 6.19   Impact analysis of energy storage units in tie-line power

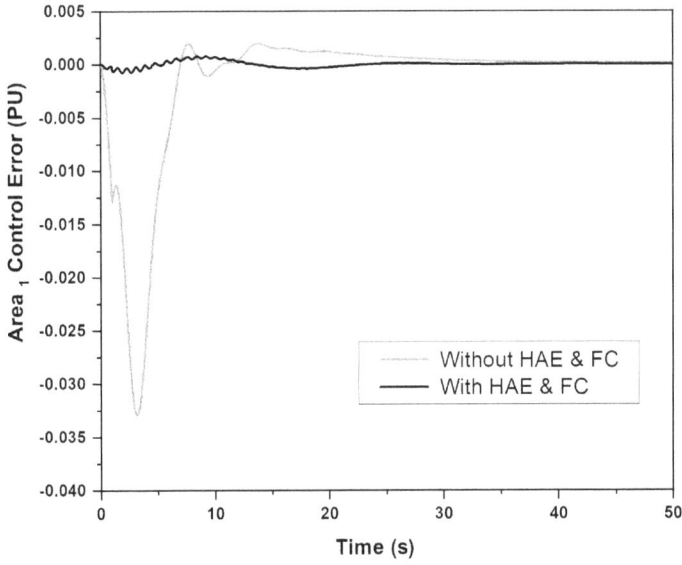

Figure 6.20   Impact analysis of energy storage units in ACE 1

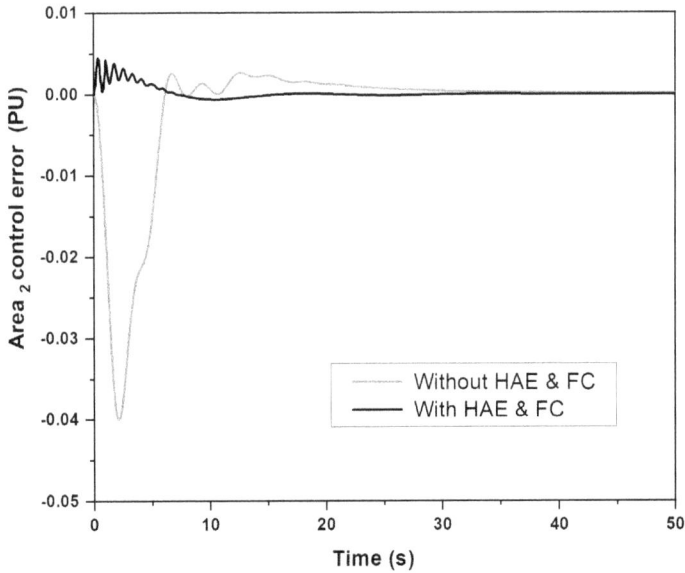

Figure 6.21   Impact analysis of energy storage units in ACE 2

optimization techniques. During the optimization, the system is under 1% load demand.

By the deep analysis of the conclusion derived in this chapter, the proposed ACO–PID controller provides an improved response and maintains system stability at the time of uncertainty loading. In the previous chapter, the system with renewable energy sources was investigated; by comparing the results, the impact of the energy storage units is understood in this chapter. The ESUs provide noticeable support to the secondary controller to maintain system stability and control the deviation in frequency. The time domain specification of the chapter provides a clear dominance of ACO over other optimization methods in terms of fast settling of system frequency oscillation and minimal peak values. ACO gives a quick settling response (F1= 47 s, f2=36 s, tie-line power =36 s, ACE1=45 s, and ACE2=42 s).

7 Design of secondary controllers for nonlinear multi-source power systems with energy storage unit and renewable energy

Frequency stability analysis for a multiarea multisource power system with renewable energy sources, energy storage units, and nonlinear components (boiler dynamics). In the previous chapter, the system without a boiler dynamics was analysed. In this chapter, a boiler dynamics unit is added to the thermal power plant as a nonlinear component with the previous mathematical model. This chapter is structured as follows: Section 7.1 focuses on the nonlinear component incorporated with the thermal power system. Section '7.2 presents a comparison of the simulation results, and Section 7.3 presents the conclusions of the chapter.

7.1 DESIGN OF A NONLINEAR POWER SYSTEM

A nonlinear component is incorporated with a renewable energy source-based power system. The nonlinear component is a boiler dynamics, which is added to the thermal power plant. One of the most important parts of a thermal power plant is the boiler, which heats water to produce steam. After that, the turbines attached to the generators are powered by this steam to create energy. The boiler is the central component of the power plant, producing steam because the chemical energy of fuel (such as coal, natural gas, or oil) is converted into thermal energy. The Simulink model of the boiler dynamics is shown in Figure 7.1.

The key components of the boiler:

- **Fuel supply:** supplies the fuel for combustion (coal, natural gas, etc.).
- **Combustion chamber:** where heat is produced by burning fuel.
- **Superheater:** The steam temperature is increased over the saturation temperature.
- **Economizer:** Before the feedwater enters the boiler, it is warmed.

DOI: 10.1201/9781003661153-7

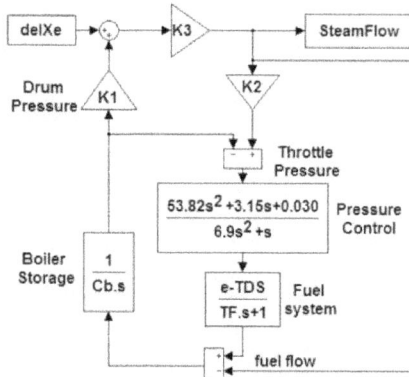

Figure 7.1   Simulink model of boiler dynamics

- **Steam drum:** Appropriate water levels are maintained by separating steam from water.
- **Air preheater:** Heats the combustion air that enters.

It is necessary to formulate a combustion reaction to model the combustion process. This entails simulating the fuel's stoichiometric air–fuel ratio and composition. For coal combustion, a simplified reaction may look like:

$$C_xH_yO_z + (x+y/4+z/2)O_2 ßxCO_2 + y/2H_2O \qquad (7.1)$$

where $C_x\,H_y\,O_z$ represents the chemical formula of the fuel.

7.2   SIMULATION RESULTS AND DISCUSSION

The ACO–PID controller response over frequency oscillation for a nonlinear (boiler dynamics) power system with energy storage units (HAE and FC) and wind energy is illustrated in this section. The PID controller design and objective function utilized in this book are explained in Chapter 2, Section 4.2. The different optimization methods (GA, PSO, FA, & BAT)-tuned PID controller gain parameters are presented in Table 7.1. A graphical comparison of the fitness values is shown in Figure 7.2.

7.2.1   RESPONSE ANALYSIS

This study utilized various optimization techniques (such as the genetic algorithm, particle swarm optimization, bat algorithm, firefly algorithm, and ant colony optimization) to adjust the controller gain parameters in the power system under investigation. The main objective was to highlight the superior performance of the ant colony optimization method in mitigating frequency oscillations. To evaluate the effectiveness of the optimized controller parameters, a 1% load demand scenario

Table 7.1

Gain values of GA-tuned, PSO-tuned, FA-tuned, BA-tuned, and ACO technique-tuned PID controller gain values with a fitness value

Technique	Kp1	Ki1	Kd1	Kp2	Ki2	Kd2	Fitness Value J
GA	0.9171	0.9174	0.5308	0.7791	0.9339	0.1300	10.7949
PSO	0.7156	0.9997	0.5888	0.9169	0.9911	0.4614	11.1997
FA	0.9999	0.9999	0.4250	0.6997	0.9893	0.1943	9.9978
BAT	0.8641	0.9958	0.3462	0.5350	0.8919	0.1498	10.4415
ACO	0.9942	0.9999	0.3514	0.6706	0.9041	0.5391	10.9295

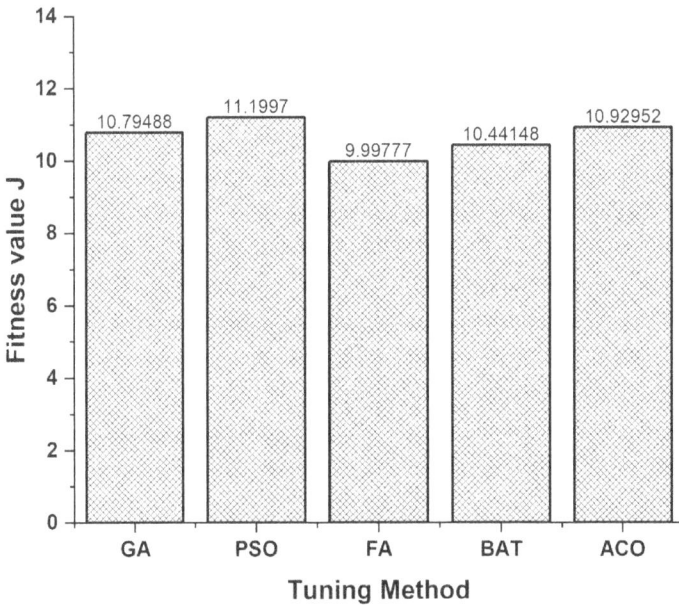

Figure 7.2 Bar chart of fitness value J

was applied. This allowed for a detailed analysis of the system's response to the changes in load demand. The results of the analysis are presented through graphical comparisons, with Figures 7.3 and 7.4 illustrating the frequency differences between area 1 and area 2. Additionally, Figure 7.5 displays the differences in tie-line power, whereas Figures 7.6 and 7.7 show the variations in the ACE values. Furthermore, the numerical data extracted from Figures 7.3 to 7.7 were compiled into Tables 7.2 to 7.4 for a more comprehensive and detailed presentation of the results. This approach allowed for a clear and systematic comparison of the performance of the different optimization techniques in addressing frequency oscillations in the power system.

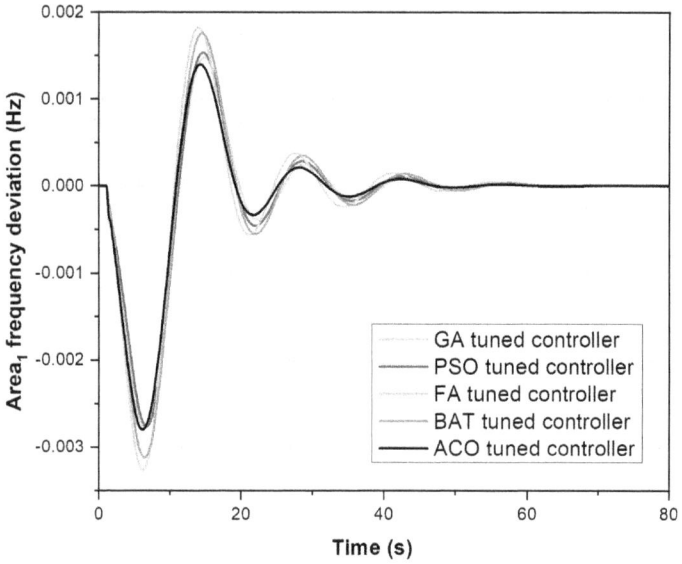

Figure 7.3   Assessment of frequency deviations in area 1

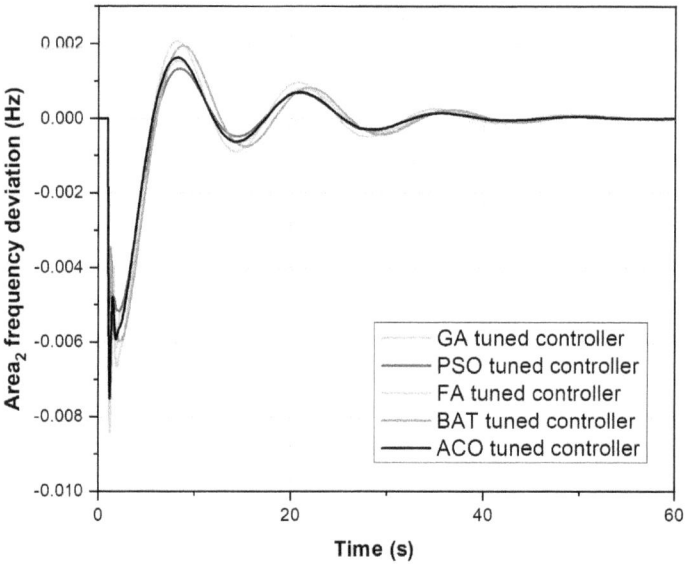

Figure 7.4   Assessment of frequency deviations in area 2

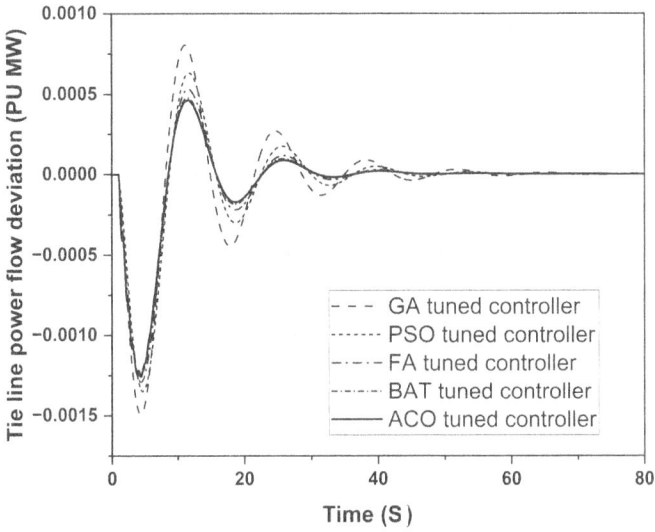

Figure 7.5   Assessment of Tie-line power flow deviations

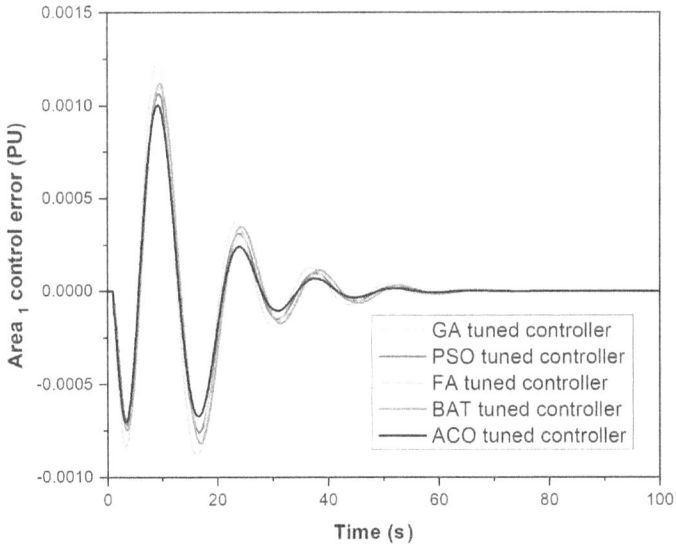

Figure 7.6   Assessment of the ACE in area 1

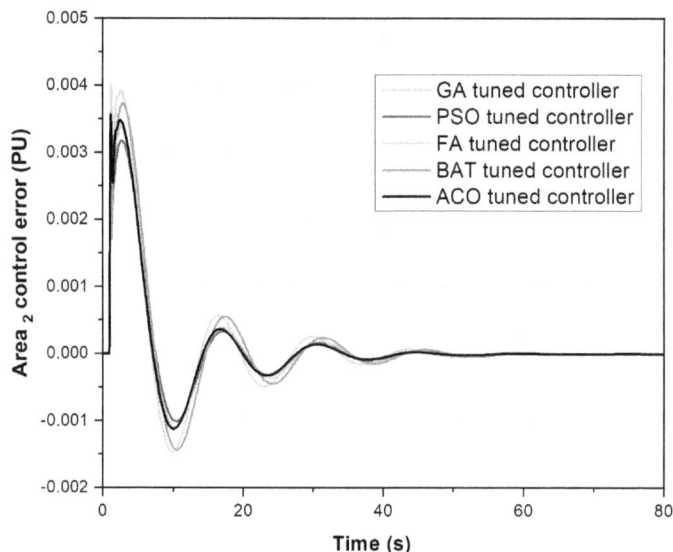

Figure 7.7   Assessment of the ACE in area 2

The comprehensive analysis of the responses in Figures 7.3 to 7.7 demonstrates the superior performance of the proposed ACO–PID controller compared with the other optimization methods employed in the study, including the GA, PSO, FA, and BAT. The ACO–PID controller effectively mitigated oscillations in the system frequency within area 1 and area 2 and the tie-line power flow within 60 s, 53 s, and 57 s, respectively, during unexpected loading scenarios. Furthermore, the ACO–PID controller exhibited quicker responses in controlling the area control error than alternative techniques did, resolving ACE 1 and ACE 2 oscillations at 46 s and 43 s. In addition to its fast-settling time, the ACO–PID controller also demonstrated a lower peak overshoot (Os) and undershoot (Us) than the other optimization methods did. A detailed numerical comparison of the aforementioned response is presented in Section '7.2.2.

7.2.2   TIME DOMAIN PARAMETERS OF THE SYSTEM WITH DIFFERENT OP-TIMIZATION TECHNIQUES TUNED WITH THE PID CONTROLLER

The mathematical values of the time domain specification parameters are listed in Tables 7.2, 7.3, and 7.4 for the system frequency in areas 1 and 2, del Ptie, and ACE in areas 1 and 2, respectively.

The performance comparisons in the previous section and the mathematical analysis in Tables 7.2, 7.3, and 7.4 above show that the ACO–PID controller settles the response in a shorter amount of time (f1 = 60 s, f2 = 53 s, tie-line = 57 s, ACE1 = 63 s, ACE2 = 58 s). Compared with the GA, PSO, FA, and BAT

Table 7.2

Optimized parameters for frequency deviations (del F1 and del F2)

Technique	del F1			del F2		
	TS (S)	$OS \times 10^{-3}(Hz)$	$US \times 10^{-3}(Hz)$	TS (S)	$OS \times 10^{-3}(Hz)$	$US \times 10^{-3}(Hz)$
GA	63	1.477	2.775	58	1.542	8.395
PSO	62	1.534	2.757	58	1.339	5.168
FA	62	1.831	3.266	58	2.079	8.452
BAT	61	1.761	3.122	56	1.953	5.968
ACO	60	1.402	2.797	53	1.636	7.512

Table 7.3

Optimized parameters for tieline power deviation (del P_{tie})

Technique	TS (S)	OS (pu MW)	US (pu MW)
GA	63	8.261×10^{-4}	1.574×10^{-3}
PSO	62	7.565×10^{-4}	1.476×10^{-3}
FA	61	9.943×10^{-4}	1.788×10^{-3}
BAT	60	9.570×10^{-4}	1.711×10^{-3}
ACO	57	7.802×10^{-4}	1.568×10^{-3}

Table 7.4

Evaluation of the time-domain parameters in the area control error

Technique	ACE 1			ACE 2		
	TS (S)	$OS \times 10^{-3}(pu)$	$US \times 10^{-4}(pu)$	TS (S)	$OS \times 10^{-3}(pu)$	$US \times 10^{-3}(pu)$
GA	65	1.102	7.788	57	3.994	1.098
PSO	69	1.005	7.608	58	3.179	1.012
FA	66	1.233	8.896	66	4.027	1.462
BAT	65	1.121	8.159	60	3.733	1.440
ACO	63	1.005	7.054	58	3.565	1.223

approach-tuned controllers, the ACO–PID controller was able to settle the oscillation faster. Figures 7.8 to 7.12 clearly illustrate the superiority of the ACO approach by plotting a PID controller bar chart for various technique-tuned controller settling times. The system frequency oscillation settles more quickly than before when energy storage devices are added because of the system's complexity in certain cases.

In Figures 7.8 to 7.12, a bar chart assessment of the settling time in areas 1 and 2, tie-line power flow, and ACE deviations demonstrated that the ACO–PID controller

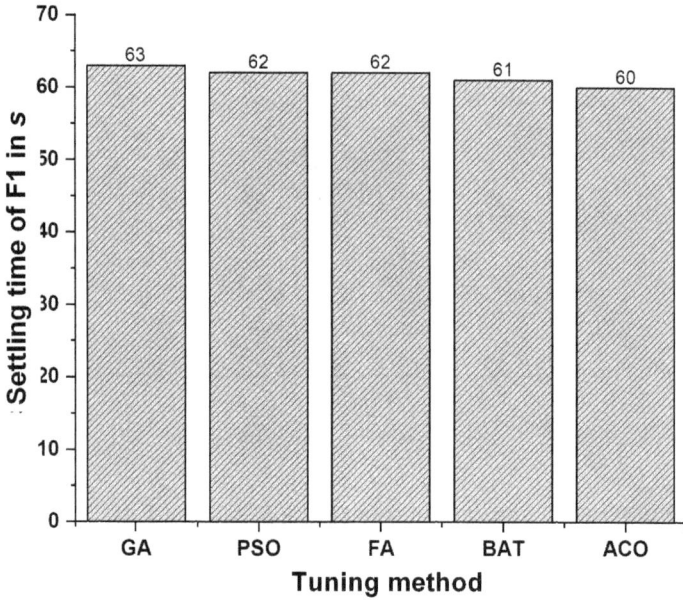

Figure 7.8   Bar chart evaluation of the settling time at the system frequency in area 1

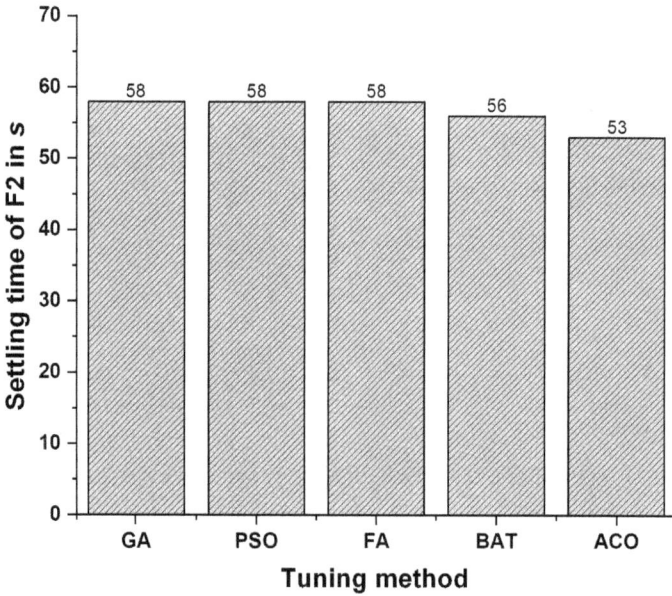

Figure 7.9   Bar chart evaluation of the settling time at the system frequency in area 2

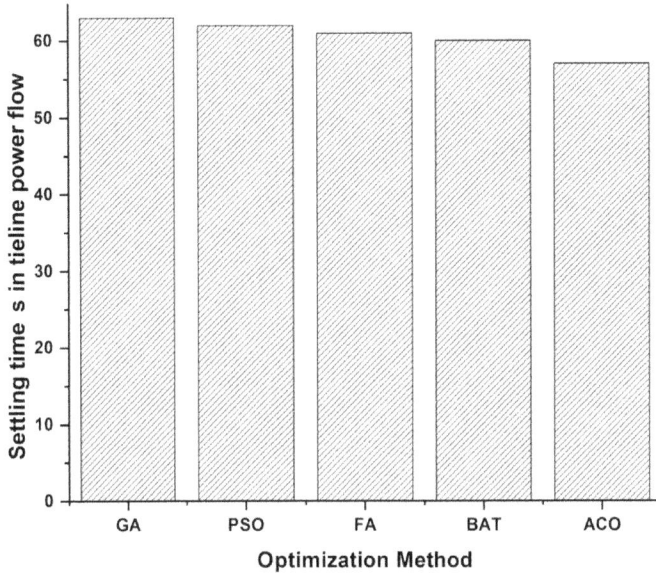

Figure 7.10   Bar chart evaluation of the tie-line power settling time

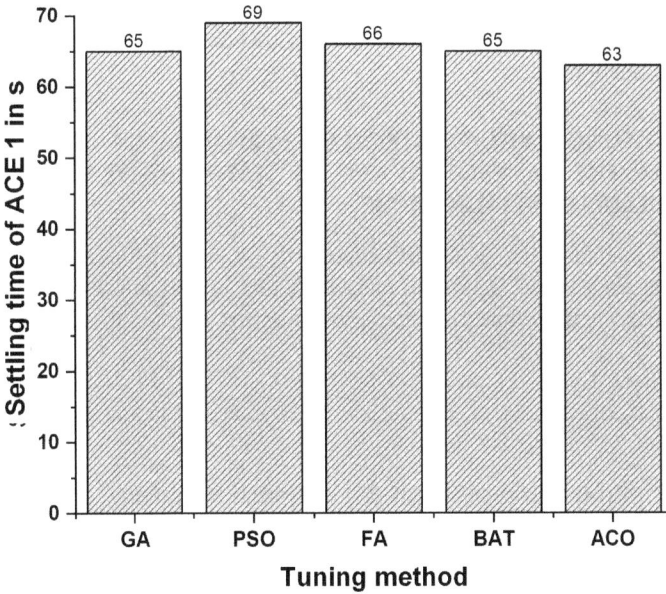

Figure 7.11   Evaluation of the settling time via a bar chart in ACE at area 1

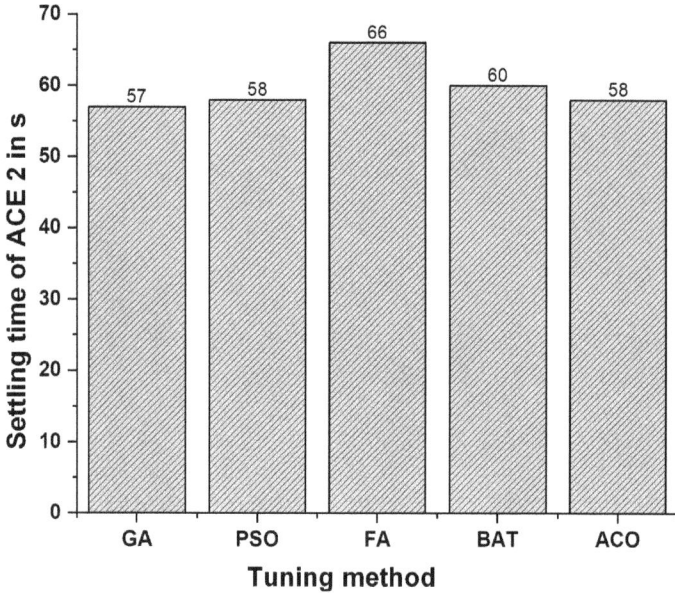

Figure 7.12   Evaluation of the settling time via a bar chart in ACE at area 2

performed better than the other optimization methods, such as the GA, PSO, FA, and BAT. The improvement analysis of ACO over other optimization techniques is explained in Section 7.2.3.

7.2.3   ACO-TUNED PID CONTROLLER PERFORMANCE IMPROVEMENT OVER THE GA, PSO, FA, AND BA TECHNIQUE-TUNED PID CONTROLLERS IN SETTLING TIME

This section evaluates the ACO–PID superiority in teams of % over other strategies, such as the genetic algorithm, particle swarm optimization technique, firefly algorithm, and bat algorithm, with numerical numbers given in the table and figures below.

Table 7.5
ACO–PID controller improvement (%) in system frequency

Technique	Area 1 System Frequency (%)	Area 2 System Frequency (%)
ACO over GA	4.7	9.4
ACO over PSO	3.1	9.4
ACO over FA	3.1	9.4
ACO over BA	1.5	5.6

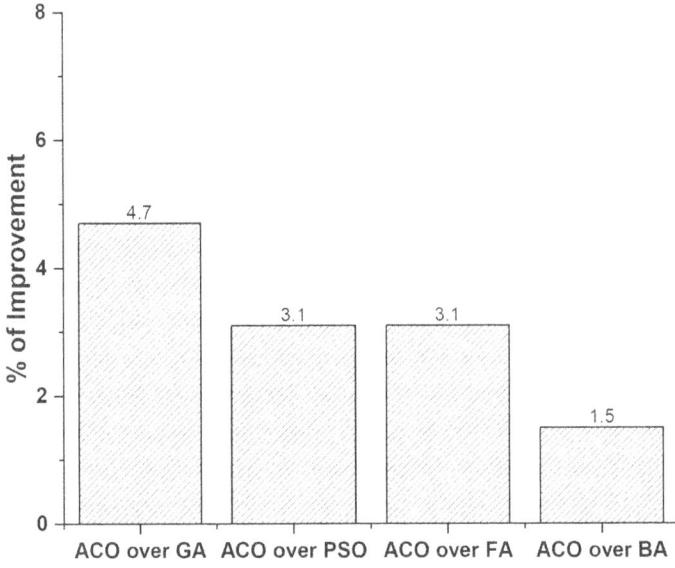

Figure 7.13   Evaluation of the settling time via a bar chart of the system frequency in area 1

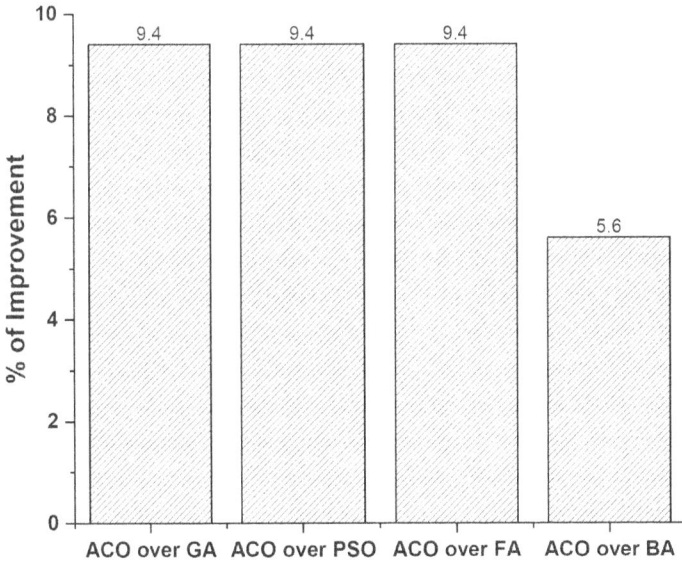

Figure 7.14   Evaluation of the settling time via a bar chart of the system frequency in area 2

Table 7.6

ACO–PID controller improvement (%) in tieline power flow

Technique	Improvement (%)
ACO over GA	10.5
ACO over PSO	8.7
ACO over FA	7.0
ACO over BA	5.2

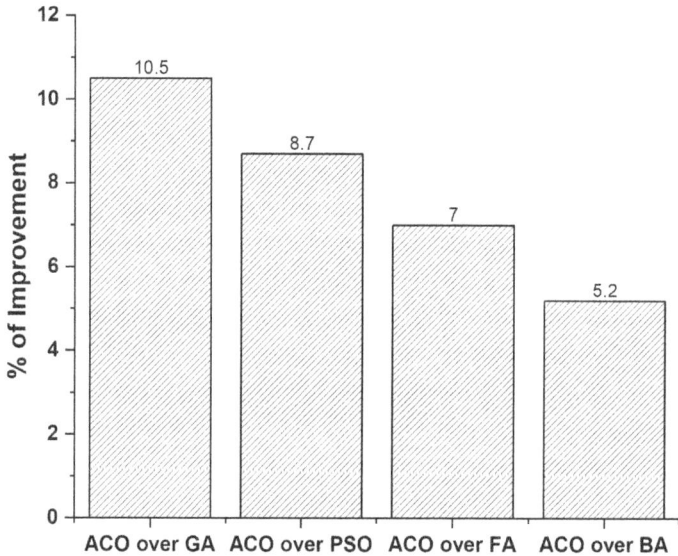

Figure 7.15   Evaluation of settling time via a bar chart for tie-line power flow

The performance enhancement analysis in Table 7.5 and the bar chart evolution in Figures 7.13 and 7.14 effectively demonstrate that the ACO–PID controller provides a more well-controlled response in terms of negligible settling and requires a longer time than the other controllers do.

The numerical values are listed in Table 7.6, and the bar chart evolution is shown in Figure 7.15, which is based on the performance enhancement analysis of the ACO-tuned controller action. Clearly, the ACO-tuned controller clearly offers a better-controlled response in terms of minimal settling, which takes more time than the other controllers do.

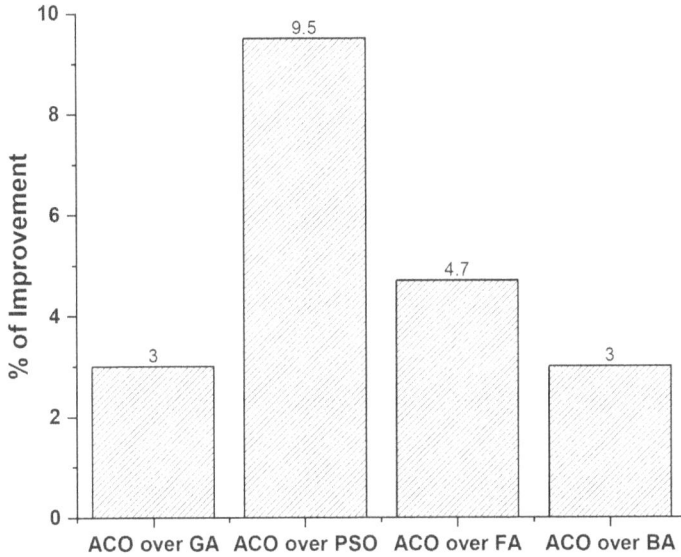

Figure 7.16   Evaluation of the settling time via a bar chart of the area control error in area 1

Table 7.7

ACO–PID controller improvement (%) in ace of area 1 and area 2

Technique	Improvement in Area 1 ACE (%)	Improvement in Area 2 ACE (%)
ACO over GA	3.0	−1.7
ACO over PSO	9.5	0.0
ACO over FA	4.7	13.7
ACO over BA	3.0	3.4

The superiority of the ACO–PID controller in terms of the numerical values of improvement is tabulated in Table 7.7, and the bar chart evolution in Figures 7.16 and 7.17 clearly shows that the ACO-tuned controller delivers a healthier controlled response.

7.2.4   IMPACT ANALYSIS OF THE NONLINEAR COMPONENT (BOILER DY-NAMICS) IN A MULTISOURCE MULTIAREA POWER SYSTEM

While energy storage units with renewable energy-based multiarea multisource power systems are included, stability issues such as settling time and peak shoot changes are controlled by the backup power from the energy storage units. The exact impact of energy storage is analysed in this section. The graphical analysis of the

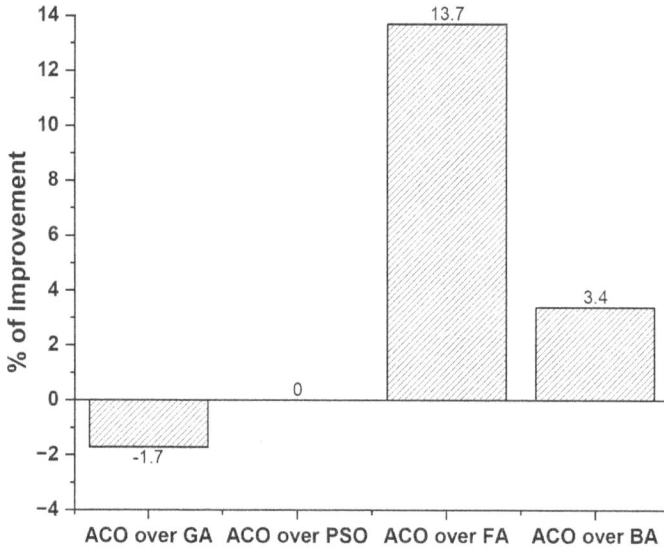

Figure 7.17 Evaluation of the settling time via a bar chart of the area control error in area 2

area 1&2 frequency difference, tie-line power, and ACE of areas 1 and 2 is shown in Figures 7.18 to 7.22. The numerical values from Figures 7.18 to 7.22 are given in Table 7.7.

Table 7.8

Comparison of optimized parameters with and without boiler dynamics

Optimized parameters/System		TS (S)	OS (Hz)	US (Hz)
Area 1 Frequency	With Boiler Dynamics	60	1.402×10^{-3}	2.797×10^{-3}
	Without Boiler Dynamics	47	9.598×10^{-4}	2.703×10^{-3}
Area 2 Frequency	With Boiler Dynamics	53	1.636×10^{-3}	7.512×10^{-3}
	Without Boiler Dynamics	36	9.636×10^{-4}	9.473×10^{-3}
Optimized parameters/System		**TS (S)**	**OS (pu MW)**	**US (pu MW)**
Tie-line power line	With Boiler Dynamics	57	7.802×10^{-4}	1.568×10^{-3}
	Without Boiler Dynamics	44	4.643×10^{-4}	1.257×10^{-3}
Optimized parameters/System		**TS (S)**	**OS (pu)**	**US (pu)**
ACE 1	With Boiler Dynamics	63	1.005×10^{-3}	7.054×10^{-4}
	Without Boiler Dynamics	46	7.751×10^{-4}	7.600×10^{-4}
ACE 2	With Boiler Dynamics	58	3.565×10^{-3}	1.223×10^{-3}
	Without Boiler Dynamics	43	4.510×10^{-3}	6.342×10^{-4}

This section analyses the impact of the nonlinear component (boiler dynamics) in a multisource multiarea with a renewable energy source power system for LFC. Figures 7.17 to 7.22 and Table 7.8 demonstrate that the system frequency oscillation

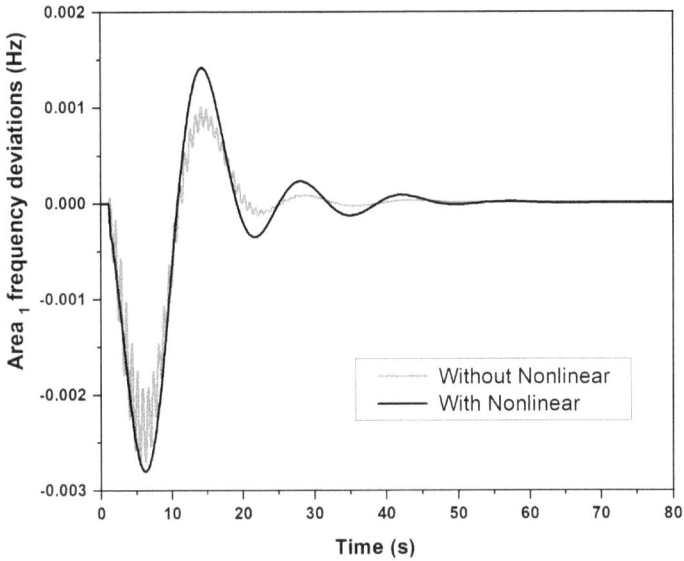

Figure 7.18 Impact analysis of the nonlinear component (boiler dynamics) in the area 1 frequency

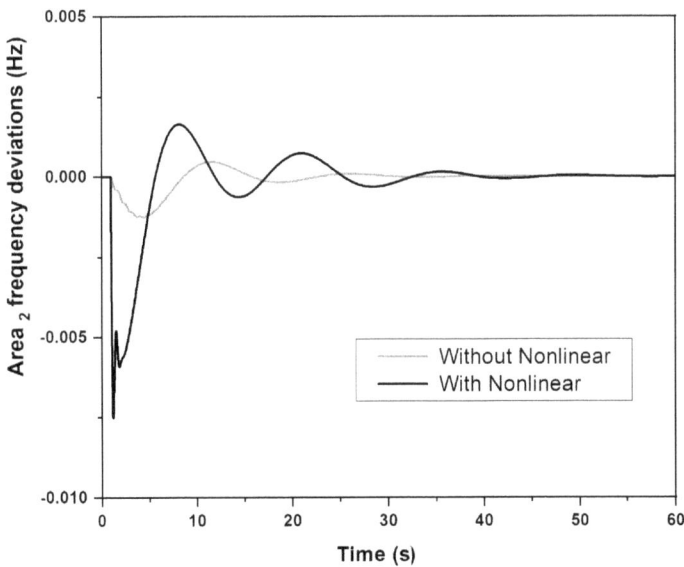

Figure 7.19 Impact analysis of the nonlinear component (boiler dynamics) in the area 2 frequency

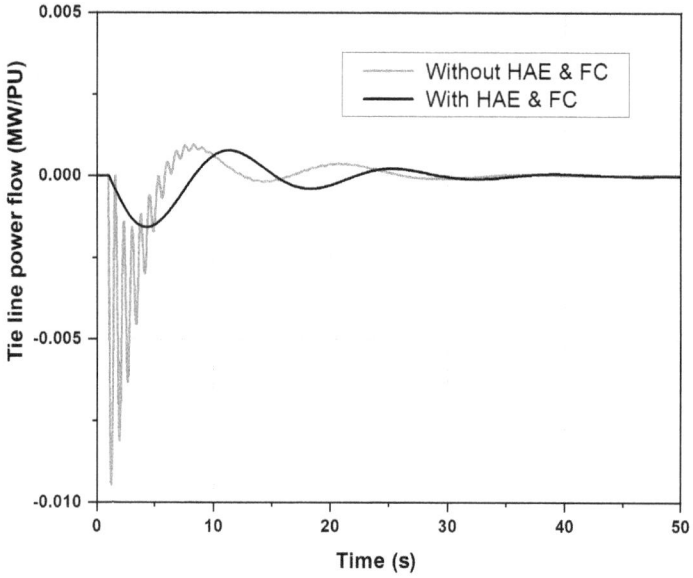

Figure 7.20   Impact analysis of the nonlinear component (boiler dynamics) of tie-line power

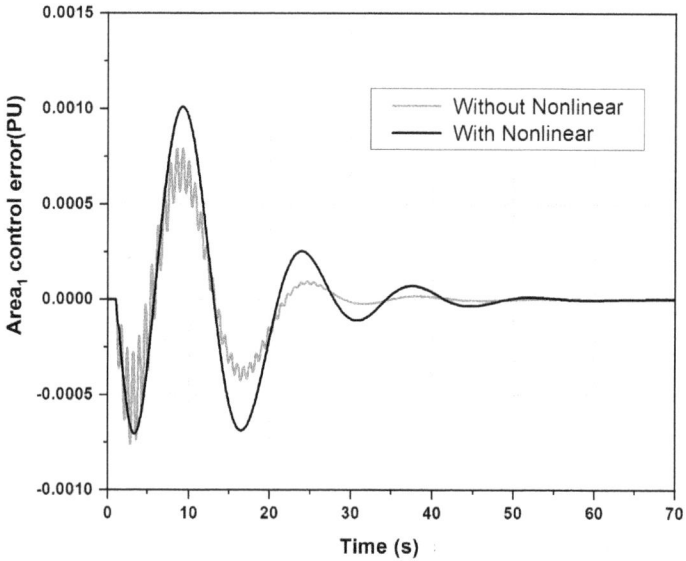

Figure 7.21   Impact analysis of the nonlinear component (boiler dynamics) in ACE 1

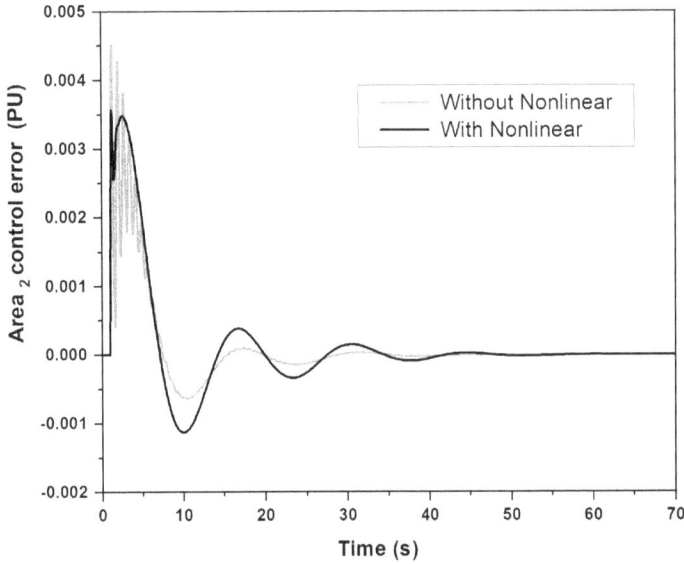

Figure 7.22 Impact analysis of the nonlinear component (boiler dynamics) in ACE 2

is disturbed when the boiler dynamics unit is integrated with the proposed power system because of the nonlinear properties of the boiler dynamics.

7.3   CONCLUSION

The proposed system includes nonlinear components (boiler dynamics) and renewable energy sources (wind) in a multisource multiarea with energy storage units. This chapter discusses the boiler dynamics impact on the thermal power plant for frequency oscillation. However, a boiler dynamics with a thermal power plant system frequency is much more disturbed than a system without a boiler dynamics unit. However, the proposed controller settles the frequency damping more quickly than the other methods do. For supremacy of the proposed method, the results of the GA, PSO, FA, and BAT are compared with those of the ACO–PID controller. The proposed ACO–PID controller provides an improved response and maintains system stability at the time of uncertainty loading. The time domain specification of the chapter provides a clear dominance of ACO over other optimization methods in terms of fast settling of system frequency oscillation and minimal peak values. ACO gives a quick settling response (F1= 60 s, f2=53 s, tie-line power =57 s, ACE1=63 s, and ACE2=58 s).

8 Challenges and future work

Recently, owing to technological advancements, the load demand has been increasing daily with different load patterns. To address this issue, several load frequency control and automatic generation control schemes have been implemented by many researchers via modern tools and techniques in single/multiarea interconnected systems or microgrid systems.

During the investigation, different load patterns with various time intervals are considered for analysis. This proposed book presents an investigation of a multiarea, multisource interconnected power system. Various optimization techniques for tuning controllers, the effects of energy storage units, nonlinearity, and renewable energy sources are applied during system investigations.

The performance of the multiarea, multisource power system is enhanced by applying new computational intelligence techniques for tuning controller gain values, ensuring that system parameters remain within predetermined limits during emergency load demand situations. Additionally, the system is equipped with energy storage units and renewable energy sources to balance the load demand. Nonlinear components are also considered in the system to validate the behavior of the proposed method. The proposed system comprises thermal, hydro, and gas generating units for investigation in a multiarea, multisource power system.

An efficient and effective control scheme is required to maintain system stability and parameters within specified limits by providing suitable control signals with the support of computational intelligence (CI) technique-tuned controllers. This investigation provides real-time support to power system engineers in balancing emergency load demand and maintaining system stability and performance within predetermined limits.

Furthermore, this research can be extended by considering the impact of electric vehicle adoption for charging/delivering power to the system to maintain stability and ensure high-quality power during emergency load demand situations. This study can be further continued by analyzing different combinations of controllers, such as PID, FOPID, and DOF controllers, in a deregulated system environment.

DOI: 10.1201/9781003661153-8

References

1. Ahmadreza Abazari, Mehdi Ghazavi Dozein, and Hassan Monsef. An optimal fuzzy-logic based frequency control strategy in a high wind penetrated power system. *Journal of the Franklin Institute*, 355(14):6262–6285, September 2018.
2. Ahmadreza Abazari, Hassan Monsef, and Bin Wu. Load frequency control by de-loaded wind farm using the optimal fuzzy-based pid droop controller. *IET Renewable Power Generation*, 13(1):180–190, October 2018.
3. A.Y. Abdelaziz and E.S. Ali. Cuckoo search algorithm based load frequency controller design for nonlinear interconnected power system. *International Journal of Electrical Power & Energy Systems*, 73:632–643, December 2015.
4. Farag K. Abo-Elyousr. Load frequency controller design for two area interconnected power system with DFIG based wind turbine via ant colony algorithm. In *2016 Eighteenth International Middle East Power Systems Conference (MEPCON)*, page 253–260. IEEE, December 2016.
5. S. Ajithapriyadarsini, P. Melba Mary, and M. Willjuice Iruthayarajan. Automatic generation control of a multi-area power system with renewable energy source under deregulated environment: adaptive fuzzy logic-based differential evolution (DE) algorithm. *Soft Computing*, 23(22):12087–12101, January 2019.
6. E.S. Ali and S.M. Abd-Elazim. BFOA based design of PID controller for two area load frequency control with nonlinearities. *International Journal of Electrical Power & Energy Systems*, 51:224–231, October 2013.
7. Anuoluwapo O. Aluko, David G. Dorrell, R Pillay-Carpanen, and Evans E. Ojo. Frequency control of modern multi-area power systems using fuzzy logic controller. In *2019 IEEE PES/IAS PowerAfrica*, page 645–649. IEEE, aug 2019.
8. B Anand and A Ebenezer Jeyakumar. Load frequency control with fuzzy logic controller considering non-linearities and boiler dynamics. *ICgST-ACSE Journal*, 8(111):15–20, 2009.
9. Yogendra Arya. Agc performance enrichment of multi-source hydrothermal gas power systems using new optimized fofpid controller and redox flow batteries. *Energy*, 127:704–715, may 2017.
10. Yogendra Arya. Agc of two-area electric power systems using optimized fuzzy pid with filter plus double integral controller. *Journal of the Franklin Institute*, 355(11):4583–4617, July 2018.
11. Yogendra Arya and Narendra Kumar. Agc of a multi-area multi-source hydrothermal power system interconnected via ac/dc parallel links under deregulated environment. *International Journal of Electrical Power & Energy Systems*, 75:127–138, February 2016.
12. Yogendra Arya, Narendra Kumar, and Ibraheem. Agc of a two-area multi-source power system interconnected via ac/dc parallel links under restructured power environment. *Optimal Control Applications and Methods*, 37(4):590–607, June 2015.
13. Flah Aymen, Naoui Mohamed, Saad Chayma, C. H. Rami Reddy, Mosleh M. Alharthi, and Sherif S. M. Ghoneim. An improved direct torque control topology of a double stator machine using the fuzzy logic controller. *IEEE Access*, 9:126400–126413, 2021.

14. B Prakash Ayyappan. Load frequency control of a multi-area power system using ann based fuzzy inference system. *International Electrical Engineering Journal (IEEJ)*, 5(6):1437–1443, 2014.

15. C. R. Balamurugan. Three area power system load frequency control using fuzzy logic controller. *International Journal of Applied Power Engineering (IJAPE)*, 7(1):18, April 2018.

16. Ajit Kumar Barisal and Deepak Kumar Lal. Application of moth flame optimization algorithm for agc of multi-area interconnected power systems. *International Journal of Energy Optimization and Engineering*, 7(1):22–49, January 2018.

17. A.K. Barisal and Somanath Mishra. Improved pso based automatic generation control of multi-source nonlinear power systems interconnected by ac/dc links. *Cogent Engineering*, 5(1):1422228, January 2018.

18. Alberto Berzoy, Johnny Rengifo, and Osama Mohammed. Fuzzy predictive dtc of induction machines with reduced torque ripple and high-performance operation. *IEEE Transactions on Power Electronics*, 33(3):2580–2587, March 2018.

19. Taieb Bessaad, Rachid Taleb, Fayçal Chabni, and Atif Iqbal. Fuzzy adaptive control of a multimachine system with single inverter supply. *International Transactions on Electrical Energy Systems*, 29(10), June 2019.

20. Ertugrul Cam, Goksu Gorel, and Hayati Mamur. Use of the genetic algorithm-based fuzzy logic controller for load-frequency control in a two area interconnected power system. *Applied Sciences*, 7(3):308, mar 2017.

21. S. Chaine, M. Tripathy, and S. Satpathy. Nsga-ii based optimal control scheme of wind thermal power system for improvement of frequency regulation characteristics. *Ain Shams Engineering Journal*, 6(3):851–863, September 2015.

22. Sabita Chaine and M. Tripathy. Design of an optimal smes for automatic generation control of two-area thermal power system using cuckoo search algorithm. *Journal of Electrical Systems and Information Technology*, 2(1):1–13, May 2015.

23. Shamik Chatterjee, Chandan Kumar Shiva, and Vivekananda Mukherjee. Automatic generation control of multi-area hydro power system using moth flame optimization technique. In *2019 3rd International Conference on Recent Developments in Control, Automation & Power Engineering (RDCAPE)*, page 395–403. IEEE, October 2019.

24. D.K. Chaturvedi, Rahul Umrao, and O.P. Malik. Adaptive polar fuzzy logic based load frequency controller. *International Journal of Electrical Power & Energy Systems*, 66:154–159, March 2015.

25. Martin Joy Cheerangal, Amit Kumar Jain, and Anandarup Das. Control of rotor field-oriented induction motor drive during input supply voltage sag. *IEEE Journal of Emerging and Selected Topics in Power Electronics*, 9(3):2789–2796, June 2021.

26. Hanbing Dan, Peng Zeng, Wenjing Xiong, Meng Wen, Mei Su, and Marco Rivera. Model predictive control-based direct torque control for matrix converter-fed induction motor with reduced torque ripple. *CES Transactions on Electrical Machines and Systems*, 5(2):90–99, June 2021.

27. D. Das, M.L. Kothari, D.P. Kothari, and J. Nanda. Variable structure control strategy to automatic generation control of interconnected reheat thermal system. *IEE Proceedings D Control Theory and Applications*, 138(6):579, 1991.

28. Puja Dash, Lalit Chandra Saikia, and Nidul Sinha. Automatic generation control of multi area thermal system using bat algorithm optimized PD–PID cascade controller. *International Journal of Electrical Power & Energy Systems*, 68:364–372, June 2015.

29. Puja Dash, Lalit Chandra Saikia, and Nidul Sinha. Comparison of performances of several facts devices using cuckoo search algorithm optimized 2DOF controllers in multi-area agc. *International Journal of Electrical Power & Energy Systems*, 65:316–324, February 2015.

30. D. de Almeida Souza, W.C.P. de Aragao Filho, and G.C.D. Sousa. Adaptive fuzzy controller for efficiency optimization of induction motors. *IEEE Transactions on Industrial Electronics*, 54(4):2157–2164, August 2007.

31. Manoj Kumar Debnath, Ranjan Kumar Mallick, and Binod Kumar Sahu. Application of hybrid differential evolution–grey wolf optimization algorithm for automatic generation control of a multi-source interconnected power system using optimal fuzzy–PID controller. *Electric Power Components and Systems*, 45(19):2104–2117, November 2017.

32. Olle Ingemar Elgerd. *Electric energy systems theory: an introduction*. McGraw-Hill Book Company, New York, NY, 1982.

33. M. Elsisi, M. Soliman, M.A.S. Aboelela, and W. Mansour. Optimal design of model predictive control with superconducting magnetic energy storage for load frequency control of nonlinear hydrothermal power system using bat inspired algorithm. *Journal of Energy Storage*, 12:311–318, August 2017.

34. Ahmed Elzawawy, Mahmoud Ali, Fahmy Bendary, and Wagdy Mansour. Adaptive under frequency load shedding scheme using genetic algorithm based artificial neural network. *International Journal of Computers*, 4, 2019.

35. Rajendra Fagna. Load frequency control of single area thermal power plant using type 1 fuzzy logic controller. *Science Journal of Circuits, Systems and Signal Processing*, 6(6):50, 2017.

36. Nabil Farah, Md. Hairul Nizam Talib, Zulkifilie Ibrahim, Qazwan Abdullah, Omer Aydogdu, Maaspaliza Azri, Jurifa Binti Mat Lazi, and Zainuddin Mat Isa. Investigation of the computational burden effects of self-tuning fuzzy logic speed controller of induction motor drives with different rules sizes. *IEEE Access*, 9:155443–155456, 2021.

37. Nabil Farah, Md. Hairul Nizam Talib, Nor Shahida Mohd Shah, Qazwan Abdullah, Zulkifilie Ibrahim, Jurifa Binti Mat Lazi, and Auzani Jidin. A novel self-tuning fuzzy logic controller based induction motor drive system: An experimental approach. *IEEE Access*, 7:68172–68184, 2019.

38. S Farook and P Sangameswara Raju. Agc controllers to optimize lfc regulation in deregulated power system. *International Journal of Advances in Engineering & Technology*, 1(5):278, 2011.

39. Hatef Farshi and Khalil Valipour. Comparative analysis of biogeography-based optimization and fuzzy logic in load frequency control. *Majlesi Journal of Energy Management*, 5(2), 2016.

40. R. Francis and I.A. Chidambaram. Optimized PI+ load–frequency controller using bwnn approach for an interconnected reheat power system with rfb and hydrogen electrolyser units. *International Journal of Electrical Power & Energy Systems*, 67:381–392, May 2015.

41. Jim Gao and Ratnesh Jamidar. Machine learning applications for data center optimization. *Google White Paper*, 21:1–13, 2014.

42. Peter Anuoluwapo Gbadega and Akshay Kumar Saha. Load frequency control of a two-area power system with a stand-alone microgrid based on adaptive model predictive control. *IEEE Journal of Emerging and Selected Topics in Power Electronics*, 9(6):7253–7263, December 2021.

43. Ali Ghasemi-Marzbali. Multi-area multi-source automatic generation control in deregulated power system. *Energy*, 201:117667, June 2020.

44. Meysam Gheisarnejad. An effective hybrid harmony search and cuckoo optimization algorithm based fuzzy PID controller for load frequency control. *Applied Soft Computing*, 65:121–138, April 2018.

45. Sajjad Golshannavaz, Rahmat Khezri, Mostafa Esmaeeli, and Pierluigi Siano. A two-stage robust-intelligent controller design for efficient lfc based on kharitonov theorem and fuzzy logic. *Journal of Ambient Intelligence and Humanized Computing*, 9(5):1445–1454, August 2017.

46. A. H. Gomaa Haroun and Yin-Ya Li. Ant lion optimized fractional order fuzzy precompensated intelligent pid controller for frequency stabilization of interconnected multi-area power systems. *Applied System Innovation*, 2(2):17, May 2019.

47. Tulasichandra Sekhar Gorripotu, Rabindra Kumar Sahu, and Sidhartha Panda. AGC of a multi-area power system under deregulated environment using redox flow batteries and interline power flow controller. *Engineering Science and Technology, an International Journal*, 18(4):555–578, December 2015.

48. Tulasichandra Sekhar Gorripotu, Halini Samalla, Ch. Jagan Mohana Rao, Ahmad Taher Azar, and Danilo Pelusi. *TLBO Algorithm Optimized Fractional-Order PID Controller for AGC of Interconnected Power System*, page 847–855. Springer Singapore, August 2018.

49. Haluk Gozde, M. Cengiz Taplamacioglu, and İlhan Kocaarslan. Comparative performance analysis of artificial bee colony algorithm in automatic generation control for interconnected reheat thermal power system. *International Journal of Electrical Power & Energy Systems*, 42(1):167–178, November 2012.

50. Dipayan Guha, Provas Kumar Roy, and Subrata Banerjee. Study of dynamic responses of an interconnected two-area all thermal power system with governor and boiler nonlinearities using BBO. In *Proceedings of the 2015 Third International Conference on Computer, Communication, Control and Information Technology (C3IT)*, page 1–6. IEEE, February 2015.

51. Dipayan Guha, Provas Kumar Roy, and Subrata Banerjee. Application of modified biogeography based optimization in agc of an interconnected multi-unit multi-source AC-DC linked power system. *International Journal of Energy Optimization and Engineering*, 5(3):1–18, July 2016.

52. Dipayan Guha, Provas Kumar Roy, and Subrata Banerjee. Oppositional biogeography-based optimisation applied to smes and TCSC-based load frequency control with generation rate constraints and time delay. *International Journal of Power and Energy Conversion*, 7(4):391, 2016.

53. Dipayan Guha, Provas Kumar Roy, and Subrata Banerjee. Symbiotic organism search algorithm applied to load frequency control of multi-area power system. *Energy Systems*, 9(2):439–468, February 2017.

54. Nizamuddin Hakimuddin, Ibraheem Nasiruddin, Terlochan Singh Bhatti, and Yogendra Arya. Optimal automatic generation control with hydro, thermal, gas, and wind power plants in 2-area interconnected power system. *Electric Power Components and Systems*, 48(6–7):558–571, April 2020.

55. Mahammad A. Hannan, Jamal A. Ali, Azah Mohamed, Ungku Anisa Ungku Amirulddin, Nadia Mei Lin Tan, and Mohammad Nasir Uddin. Quantum-behaved lightning search algorithm to improve indirect field-oriented fuzzy-pi control for im drive. *IEEE Transactions on Industry Applications*, 54(4):3793–3805, July 2018.

56. Chongxin Huang, Dong Yue, Xiangpeng Xie, and Jun Xie. Anti-windup load frequency controller design for multi-area power system with generation rate constraint. *Energies*, 9(5):330, April 2016.

57. K. Jagatheesan and B. Anand. Dynamic performance of multi-area hydro thermal power systems with integral controller considering various performance indices methods. In *2012 International Conference on Emerging Trends in Science, Engineering and Technology (INCOSET)*, page 474–478. IEEE, December 2012.

58. K Jagatheesan and B Anand. Automatic generation control of three area hydro-thermal power systems considering electric and mechanical governor with conventional controller and ant colony optimization technique. *Advances in Natural and Applied Sciences*, 8(20):25–34, 2014.

59. K. Jagatheesan and B. Anand. Performance analysis of three area thermal power systems with different steam system configurations considering non linearity and boiler dynamics using conventional controller. In *2015 International Conference on Computer Communication and Informatics (ICCCI)*, page 1–8. IEEE, January 2015.

60. K. Jagatheesan, B. Anand, K. Baskaran, N. Dey, A.S. Ashour, and V.E. Balas. *Effect of Nonlinearity and Boiler Dynamics in Automatic Generation Control of Multi-area Thermal Power System with Proportional-Integral-Derivative and Ant Colony Optimization Technique*, page 89–110. Springer International Publishing, July 2017.

61. K. Jagatheesan, B. Anand, K. Nilanjan Dey, Amira S. Ashour, and Suresh Chandra Satapathy. Performance evaluation of objective functions in automatic generation control of thermal power system using ant colony optimization technique-designed proportional–integral–derivative controller. *Electrical Engineering*, 100(2):895–911, May 2017.

62. K. Jagatheesan, B. Anand, and Nilanjan Dey. *Evolutionary Algorithm Based LFC of Single Area Thermal Power System with Different Steam Configurations and Nonlinearity*, page 185–194. Springer India, 2016.

63. K Jagatheesan, B Anand, Nilanjan Dey, and Amira S Ashour. Artificial intelligence in performance analysis of load frequency control in thermal-wind-hydro power systems. *Artif Intell*, 6(7):203–212, 2015.

64. K. Jagatheesan, B. Anand, Nilanjan Dey, Amira S. Ashour, and Valentina E. Balas. *Load Frequency Control of Hydro-Hydro System with Fuzzy Logic Controller Considering Non-linearity*, page 307–318. Springer International Publishing, 2018.

65. K. Jagatheesan, B. Anand, Nillanjan Dey, Tarek Gaber, Aboul Ella Hassanien, and Tai-Hoon Kim. A design of pi controller using stochastic particle swarm optimization in load frequency control of thermal power systems. In *2015 Fourth International Conference on Information Science and Industrial Applications (ISI)*, page 25–32. IEEE, September 2015.

66. K Jagatheesan, B Anand, and MA Ebrahim. Stochastic particle swarm optimization for tuning of pid controller in load frequency control of single area reheat thermal power system. *Int. J. Electr. Power Eng*, 8(2):33–40, 2014.

67. K. Jagatheesan, B. Anand, Sourav Samanta, Nilanjan Dey, Amira S. Ashour, and Valentina E. Balas. Particle swarm optimisation-based parameters optimisation of pid controller for load frequency control of multi-area reheat thermal power systems. *International Journal of Advanced Intelligence Paradigms*, 9(5/6):464, 2017.

68. K. Jagatheesan, B. Anand, Sourav Samanta, Nilanjan Dey, Amira S. Ashour, and Valentina E. Balas. Design of a proportional-integral-derivative controller for an automatic generation control of multi-area power thermal systems using firefly algorithm. *IEEE/CAA Journal of Automatica Sinica*, 6(2):503–515, March 2019.

69. K. Jagatheesan, B. Anand, Soumadip Sen, and Sourav Samanta. *Application of Chaos-Based Firefly Algorithm Optimized Controller for Automatic Generation Control of Two Area Interconnected Power System with Energy Storage Unit and UPFC*, page 173–191. Springer Singapore, November 2019.

70. Jagaheesan Kaliannan and Anand Baskaran. Performance analysis of double reheat turbine in multi-area agc system using conventional and ant colony optimization technique. *IU-Journal of Electrical & Electronics Engineering*, 15(1):1849–1854, 2015.

71. Suhas Vilas Kamble and S. M. Akolkar. Load frequency control of micro hydro power plant using fuzzy logic controller. In *2017 IEEE International Conference on Power, Control, Signals and Instrumentation Engineering (ICPCSI)*, page 1783–1787. IEEE, September 2017.

72. S. Kayalvizhi and D. M. Vinod Kumar. Load frequency control of an isolated micro grid using fuzzy adaptive model predictive control. *IEEE Access*, 5:16241–16251, 2017.

73. Abbas Ketabi and Masoud Hajiakbari Fini. An adaptive set-point modulation technique to enhance the performance of load frequency controllers in a multi-area power system. *Journal of Electrical Systems and Information Technology*, 2(3):391–405, December 2015.

74. Rajendra Ku Khadanga and Jitendriya Ku Satapathy. A new hybrid GA–GSA algorithm for tuning damping controller parameters for a unified power flow controller. *International Journal of Electrical Power & Energy Systems*, 73:1060–1069, December 2015.

75. Rajendra Ku Khadanga and Jitendriya Ku Satapathy. Time delay approach for pss and sssc based coordinated controller design using hybrid PSO–GSA algorithm. *International Journal of Electrical Power & Energy Systems*, 71:262–273, October 2015.

76. Rajendra Kumar Khadanga and Amit Kumar. Analysis of pid controller for the load frequency control of static synchronous series compensator and capacitive energy storage source-based multi-area multi-source interconnected power system with hvdc link. *International Journal of Bio-Inspired Computation*, 13(2):131, 2019.

77. Saad Khadar, Haitham Abu-Rub, and Abdellah Kouzou. Sensorless field-oriented control for open-end winding five-phase induction motor with parameters estimation. *IEEE Open Journal of the Industrial Electronics Society*, 2:266–279, 2021.

78. Rahmat Khezri, Arman Oshnoei, Soroush Oshnoei, Hassan Bevrani, and S.M. Muyeen. An intelligent coordinator design for gcsc and agc in a two-area hybrid power system. *Applied Soft Computing*, 76:491–504, March 2019.

79. Dwarkadas Pralhaddas Kothari and IJ Nagrath. *Power System Engineering*. Tata McGraw-Hill New Delhi, India, 2008.

80. M.L. Kothari, J. Nanda, D.P. Kothari, and D. Das. Discrete-mode automatic generation control of a two-area reheat thermal system with new area control error. *IEEE Transactions on Power Systems*, 4(2):730–738, May 1989.

81. N El Y Kouba, A Benseddik, Y Amrane, M Hasni, and M Menaa. Coordinated control of optimal lfc method and energy storage system for microgrid frequency regulation in presence of wind farm. *International Journal of Electronic and Electrical Engineering Systems*, 1(1):9–14, 2018.

82. Rajiv Kumar and V. K. Sharma. Whale optimization controller for load frequency control of a two-area multi-source deregulated power system. *International Journal of Fuzzy Systems*, 22(1):122–137, December 2019.

83. K. Kumari, G. Shankar, S. Kumari, and S. Gupta. Load frequency control using ANN-PID controller. In *2016 IEEE 1st International Conference on Power Electronics, Intelligent Control and Energy Systems (ICPEICES)*, page 1–6. IEEE, July 2016.

84. Prabha Kundur. Power system stability. *Power System Stability and Control*, 10(1):7–1, 2007.

85. Yann LeCun, Yoshua Bengio, and Geoffrey Hinton. Deep learning. *Nature*, 521(7553): 436–444, May 2015.

86. Kangdi Lu, Wuneng Zhou, Guoqiang Zeng, and Yiyuan Zheng. Constrained population extremal optimization-based robust load frequency control of multi-area interconnected power system. *International Journal of Electrical Power & Energy Systems*, 105:249–271, February 2019.

87. T Luo, M J Hou, S L Fang, and C Y Ma. Coordinated control of agc with consideration pevs and controllable loads in power system with renewable energy sources. *IOP Conference Series: Earth and Environmental Science*, 188:012070, October 2018.

88. Mazin Mustafa Mahdi and Abu Zaharin Ahmad. Load frequency control in microgrid using fuzzy logic table control. In *2017 11th IEEE International Conference on Compatibility, Power Electronics and Power Engineering (CPE-POWERENG)*, page 318–323. IEEE, 2017.

89. Mazin Mustafa Mahdi, Ekhlas Mhawi Thajeel, and Abu Zaharin Ahmad. Load frequency control for hybrid micro-grid using MRAC with ANN under-sudden load changes. In *2018 Third Scientific Conference of Electrical Engineering (SCEE)*, page 220–225. IEEE, December 2018.

90. Akhilesh Kumar Mishra and Puneet Mishra. Improved fractional order control of a nonlinear interconnected power system using salp swarm algorithm. In *2019 IEEE 16th India Council International Conference (INDICON)*, page 1–4. IEEE, December 2019.

91. Rajendra Mistry, William R. Finley, Emam Hashish, and Scott Kreitzer. Rotating machines: The pros and cons of monitoring devices. *IEEE Industry Applications Magazine*, 24(6):44–55, November 2018.

92. Tigist Mohammed, James Momoh, and Anup Shukla. Single area load frequency control using fuzzy-tuned pi controller. In *2017 North American Power Symposium (NAPS)*, page 1–6. IEEE, September 2017.

93. Banaja Mohanty, B.V.S. Acharyulu, and P.K. Hota. Moth-flame optimization algorithm optimized dual-mode controller for multiarea hybrid sources agc system. *Optimal Control Applications and Methods*, 39(2):720–734, November 2017.

94. Banaja Mohanty and Prakash Kumar Hota. Comparative performance analysis of fruit fly optimisation algorithm for multi-area multi-source automatic generation control under deregulated environment. *IET Generation, Transmission & Distribution*, 9(14):1845–1855, November 2015.

95. Tapas Kumar Mohapatra and Binod Kumar Sahu. Design and implementation of ssa based fractional order pid controller for automatic generation control of a multi-area, multi-source interconnected power system. In *2018 Technologies for Smart-City Energy Security and Power (ICSESP)*, page 1–6. IEEE, March 2018.

96. Javad Morsali, Kazem Zare, and Mehrdad Tarafdar Hagh. Applying fractional order pid to design tcsc-based damping controller in coordination with automatic generation control of interconnected multi-source power system. *Engineering Science and Technology, an International Journal*, 20(1):1–17, February 2017.

97. Venkataramana Naik N, Aurobinda Panda, and S. P. Singh. A three-level fuzzy-2 DTC of induction motor drive using SVPWM. *IEEE Transactions on Industrial Electronics*, 63(3):1467–1479, March 2016.

98. Kanendra Naidu, Hazlie Mokhlis, Ab Halim Abu Bakar, and Vladimir Terzija. Performance investigation of abc algorithm in multi-area power system with multiple interconnected generators. *Applied Soft Computing*, 57:436–451, August 2017.

99. Venkataramana Naik N and Sajjan Pal Singh. A novel interval type-2 fuzzy-based direct torque control of induction motor drive using five-level diode-clamped inverter. *IEEE Transactions on Industrial Electronics*, 68(1):149–159, January 2021.

100. J. Nanda. Automatic generation control with fuzzy logic controller considering generation rate constraint. In *6th International Conference on Advances in Power System Control, Operation and Management. Proceedings. APSCOM 2003*, volume 2003, page 770–775. IEE, 2003.

101. J Nanda and Lalit Chandra Saikia. Comparison of performances of several types of classical controller in automatic generation control for an interconnected multi-area thermal system. In *2008 Australasian Universities Power Engineering Conference*, pages 1–6. IEEE, 2008.

102. Pratap Chandra Nayak, Abhilipsa Sahoo, Rupali Balabantaraya, and Ramesh Chandra Prusty. Comparative study of sos & pso for fuzzy based pid controller in AGC in an integrated power system. In *2018 Technologies for Smart-City Energy Security and Power (ICSESP)*, page 1–6. IEEE, March 2018.

103. Gia Nhu Nguyen, K. Jagatheesan, Amira S. Ashour, B. Anand, and Nilanjan Dey. *Ant Colony Optimization Based Load Frequency Control of Multi-area Interconnected Thermal Power System with Governor Dead-Band Nonlinearity*, page 157–167. Springer Singapore, December 2017.

104. Sasmita Padhy and Sidhartha Panda. A hybrid stochastic fractal search and pattern search technique based cascade pi-pd controller for automatic generation control of multi-source power systems in presence of plug in electric vehicles. *CAAI Transactions on Intelligence Technology*, 2(1):12–25, March 2017.

105. Nikhil Paliwal, Laxmi Srivastava, and Manjaree Pandit. Application of grey wolf optimization algorithm for load frequency control in multi-source single area power system. *Evolutionary Intelligence*, 15(1):563–584, November 2020.

106. Indranil Pan and Saptarshi Das. Fractional-order load-frequency control of interconnected power systems using chaotic multi-objective optimization. *Applied Soft Computing*, 29:328–344, April 2015.

107. Jay K. Pandit, Mohan V. Aware, Ronak V. Nemade, and Emil Levi. Direct torque control scheme for a six-phase induction motor with reduced torque ripple. *IEEE Transactions on Power Electronics*, 32(9):7118–7129, September 2017.

108. Nimai Charan Patel and Manoj Kumar Debnath. *Whale Optimization Algorithm Tuned Fuzzy Integrated PI Controller for LFC Problem in Thermal-hydro-wind Interconnected System*, page 67–77. Springer Singapore, 2019.

109. Chittaranjan Pradhan and Chandrashekhar N. Bhende. Online load frequency control in wind integrated power systems using modified Jaya optimization. *Engineering Applications of Artificial Intelligence*, 77:212–228, January 2019.

110. Asadur Rahman, Lalit C. Saikia, and Nidul Sinha. AGC of dish-stirling solar thermal integrated thermal system with biogeography based optimised three degree of freedom pid controller. *IET Renewable Power Generation*, 10(8):1161–1170, July 2016.

111. Asadur Rahman, Lalit Chandra Saikia, and Nidul Sinha. Load frequency control of a hydro-thermal system under deregulated environment using biogeography-based optimised three-degree-of-freedom integral-derivative controller. *IET Generation, Transmission & Distribution*, 9(15):2284–2293, November 2015.

112. Asadur Rahman, Lalit Chandra Saikia, and Nidul Sinha. Maiden application of hybrid pattern search-biogeography based optimisation technique in automatic generation control of a multi-area system incorporating interline power flow controller. *IET Generation, Transmission & Distribution*, 10(7):1654–1662, May 2016.

113. Utkarsh Raj and Ravi Shankar. *Effectiveness of Whale Optimization Based I+PD Controller for LFC of Plug-in Electric Vehicle Included Multi-area System*, page 11–19. Springer Singapore, 2020.

114. K. S. Rajesh and S. S. Dash. Load frequency control of autonomous power system using adaptive fuzzy based pid controller optimized on improved sine cosine algorithm. *Journal of Ambient Intelligence and Humanized Computing*, 10(6):2361–2373, May 2018.

115. T. Rajesh, B. Gunapriya, M. Sabarimuthu, S. Karthikkumar, R. Raja, and M. Karthik. Frequency control of pv-connected micro grid system using fuzzy logic controller. *Materials Today: Proceedings*, 45:2260–2264, 2021.

116. Naladi Ram Babu and Lalit Chandra Saikia. Automatic generation control of a solar thermal and dish-stirling solar thermal system integrated multi-area system incorporating accurate hvdc link model using crow search algorithm optimised fopi minus fodf controller. *IET Renewable Power Generation*, 13(12):2221–2231, July 2019.

117. Basavarajappa Sokke Rameshappa and Nagaraj Mudakapla Shadaksharappa. An optimal artificial neural network controller for load frequency control of a four-area interconnected power system. *International Journal of Electrical and Computer Engineering (IJECE)*, 12(5):4700, October 2022.

118. Basavarajappa Sokke Rameshappa and Nagaraj Mudakapla Shadaksharappa. *Enhancing Load Frequency Control in a Four-Area Power System Network with an Optimal ANN Controller*, page 149–168. B P International, January 2024.

119. Gummadi Srinivasa Rao, YP Obulesh, and M Kavya. Tuning of pid controller in multi area interconnected power system using particle swarm optimization. *Journal of Electrical and Electronics Engineering*, 10, 2015.

120. Amin Safari, Farshad Babaei, and Meisam Farrokhifar. A load frequency control using a PSO-based ANN for micro-grids in the presence of electric vehicles. *International Journal of Ambient Energy*, 42(6):688–700, January 2019.

121. Arindita Saha and Lalit Chandra Saikia. Performance analysis of combination of ultra-capacitor and superconducting magnetic energy storage in a thermal-gas AGC system with utilization of whale optimization algorithm optimized cascade controller. *Journal of Renewable and Sustainable Energy*, 10(1), January 2018.

122. Erdinc Sahin. Design of an optimized fractional high order differential feedback controller for load frequency control of a multi-area multi-source power system with non-linearity. *IEEE Access*, 8:12327–12342, 2020.

123. B.P. Sahoo and S. Panda. Improved grey wolf optimization technique for fuzzy aided pid controller design for power system frequency control. *Sustainable Energy, Grids and Networks*, 16:278–299, December 2018.

124. Rabindra Kumar Sahu, Sidhartha Panda, and Saroj Padhan. A hybrid firefly algorithm and pattern search technique for automatic generation control of multi area power systems. *International Journal of Electrical Power & Energy Systems*, 64:9–23, January 2015.

125. Rabindra Kumar Sahu, Sidhartha Panda, and Saroj Padhan. A novel hybrid gravitational search and pattern search algorithm for load frequency control of nonlinear power system. *Applied Soft Computing*, 29:310–327, April 2015.

126. Rabindra Kumar Sahu, Sidhartha Panda, and Umesh Kumar Rout. De optimized parallel 2-DOF PID controller for load frequency control of power system with governor dead-band nonlinearity. *International Journal of Electrical Power & Energy Systems*, 49:19–33, July 2013.

127. Lalit Chandra Saikia, Nidul Sinha, and J. Nanda. Maiden application of bacterial foraging based fuzzy IDD controller in agc of a multi-area hydrothermal system. *International Journal of Electrical Power & Energy Systems*, 45(1):98–106, February 2013.

128. D Sambariya and Vivek Nath. Load frequency control using fuzzy logic based controller for multi-area power system. *British Journal of Mathematics & Computer Science*, 13(5):1–19, January 2016.

129. D K Sambariya and Rajendra Fagna. A novel elephant herding optimization based pid controller design for load frequency control in power system. In *2017 International Conference on Computer, Communications and Electronics (Comptelix)*, page 595–600. IEEE, July 2017.

130. Prasun Sanki, Mousumi Basu, and Partha Sarathi Pal. Study of agc as two area thermal interconnected power system consisting WPG and SPG. In *2018 Emerging Trends in Electronic Devices and Computational Techniques (EDCT)*, page 1–6. IEEE, March 2018.

131. M.R. Sathya and M. Mohamed Thameem Ansari. Load frequency control using bat inspired algorithm based dual mode gain scheduling of pi controllers for interconnected power system. *International Journal of Electrical Power & Energy Systems*, 64:365–374, January 2015.

132. Takamasa Sato, Atsushi Umemura, Rion Takahashi, and Junji Tamura. Frequency control of power system with large scale wind farm installed by using HVDC transmission system. In *2017 IEEE Manchester PowerTech*, page 1–6. IEEE, June 2017.

133. Hamed Shabani, Behrooz Vahidi, and Majid Ebrahimpour. A robust pid controller based on imperialist competitive algorithm for load-frequency control of power systems. *ISA Transactions*, 52(1):88–95, January 2013.

134. Ravi Shankar, Ashiwani Kumar, Utkarsh Raj, and Kalyan Chatterjee. Fruit fly algorithm-based automatic generation control of multiarea interconnected power system with facts and AC/DC links in deregulated power environment. *International Transactions on Electrical Energy Systems*, 29(1):e2690, August 2018.

135. Yatin Sharma and Lalit Chandra Saikia. Automatic generation control of a multi-area ST – thermal power system using grey wolf optimizer algorithm based classical controllers. *International Journal of Electrical Power & Energy Systems*, 73:853–862, December 2015.

136. Mojtaba Shivaie, Mohammad G. Kazemi, and Mohammad T. Ameli. A modified harmony search algorithm for solving load-frequency control of non-linear interconnected hydrothermal power systems. *Sustainable Energy Technologies and Assessments*, 10:53–62, June 2015.

137. Vijay P. Singh, Nand Kishor, and Paulson Samuel. Improved load frequency control of power system using lmi based pid approach. *Journal of the Franklin Institute*, 354(15):6805–6830, October 2017.

138. Ashim Sonowal and Tilok Boruah. ANN and ANFIS controller for load frequency control and automatic voltage regulation. In *Intelligent Computation and Analytics on Sustainable Energy and Environment*, pages 27–32. CRC Press, 2024.

139. S Subha. Load frequency control with fuzzy logic controller considering governor dead band and generation rate constraint non-linearities. *World Appl Sci J*, 29(8):1059–1066, 2014.

140. H Sudheer, SF Kodad, and B Sarvesh. Regular paper improved fuzzy logic based DTC of induction machine for wide range of speed control using AI based controllers. *J. Electr. Syst*, 12(2):301–314, 2016.

141. AK Swain and AK Mohanty. Adaptive load frequency control of an interconnected hydro thermal system considering generation rate constraints. *Journal-Institution of Engineers India Part EL Electrical Engineering Division*, 76:109–114, 1995.

142. Seyed Abbas Taher, Masoud Hajiakbari Fini, and Saber Falahati Aliabadi. Fractional order PID controller design for LFC in electric power systems using imperialist competitive algorithm. *Ain Shams Engineering Journal*, 5(1):121–135, March 2014.

143. Qazwan A. Tarbosh, Omer Aydogdu, Nabil Farah, Md Hairul Nizam Talib, Adeeb Salh, Nihat Cankaya, Fuad Alhaj Omar, and Akif Durdu. Review and investigation of simplified rules fuzzy logic speed controller of high performance induction motor drives. *IEEE Access*, 8:49377–49394, 2020.

144. Zoheir Tir, Om P. Malik, and Ali M. Eltamaly. Fuzzy logic based speed control of indirect field oriented controlled double star induction motors connected in parallel to a single six-phase inverter supply. *Electric Power Systems Research*, 134:126–133, May 2016.

145. Zoheir Tir, Youcef Soufi, Mohammad Naser Hashemnia, Om P. Malik, and Khoudir Marouani. Fuzzy logic field oriented control of double star induction motor drive. *Electrical Engineering*, 99(2):495–503, May 2016.

146. Pretty Neelam Topno and Saurabh Chanana. Fractional order PID control for LFC problem of a hydro-thermal power system. In *2016 11th International Conference on Industrial and Information Systems (ICIIS)*, page 867–872. IEEE, December 2016.

147. Pretty Neelam Topno and Saurabh Chanana. Non-integer order control for LFC problem of two-area thermal power system with grc. In *2017 International Conference on Innovations in Electrical, Electronics, Instrumentation and Media Technology (ICEEIMT)*, page 87–91. IEEE, February 2017.

148. Debasis Tripathy, Amar Kumar Barik, Nalin Behari Dev Choudhury, and Binod Kumar Sahu. *Performance Comparison of SMO-Based Fuzzy PID Controller for Load Frequency Control*, page 879–892. Springer Singapore, October 2018.

149. Mehmet Rida Tur, Mohammed Wadi, Abdulfetah Shobole, and Selim Ay. Load frequency control of two area interconnected power system using fuzzy logic control and PID controller. In *2018 7th International Conference on Renewable Energy Research and Applications (ICRERA)*, page 1253–1258. IEEE, October 2018.

150. Chuan Xia and Huijia Liu. Bi-level model predictive control for optimal coordination of multi-area automatic generation control units under wind power integration. *Processes*, 7(10):669, September 2019.

151. Hassan A. Yousef, Khalfan AL-Kharusi, Mohammed H. Albadi, and Nasser Hosseinzadeh. Load frequency control of a multi-area power system: An adaptive fuzzy logic approach. *IEEE Transactions on Power Systems*, 29(4):1822–1830, jul 2014.

152. Mohamed S. Zaky and Mohamed K. Metwaly. A performance investigation of a four-switch three-phase inverter-fed in drives at low speeds using fuzzy logic and PI controllers. *IEEE Transactions on Power Electronics*, 32(5):3741–3753, May 2017.

153. Chunyu Zhang, Shouxiang Wang, and Qianyu Zhao. Distributed economic MPC for lfc of multi-area power system with wind power plants in power market environment. *International Journal of Electrical Power & Energy Systems*, 126:106548, March 2021.

154. Wei Zhang and Kailun Fang. Controlling active power of wind farms to participate in load frequency control of power systems. *IET Generation, Transmission & Distribution*, 11(9):2194–2203, June 2017.

155. Shuai Zhao, Frede Blaabjerg, and Huai Wang. An overview of artificial intelligence applications for power electronics. *IEEE Transactions on Power Electronics*, 36(4):4633–4658, April 2021.

Index

For Product Safety Concerns and Information please contact our EU
representative GPSR@taylorandfrancis.com
Taylor & Francis Verlag GmbH, Kaufingerstraße 24, 80331 München, Germany